NEW URBANISM

New Urbanism
Life, Work, and Space in the New Downtown

Edited by

ILSE HELBRECHT
Humboldt-Universität zu Berlin, Germany

PETER DIRKSMEIER
Humboldt-Universität zu Berlin, Germany

Routledge
Taylor & Francis Group

LONDON AND NEW YORK

First published 2012 by Ashgate Publishing

Published 2016 by Routledge
2 Park Square, Milton Park, Abingdon, Oxfordshire OX14 4RN
711 Third Avenue, New York, NY 10017, USA

First issued in paperback 2016

Routledge is an imprint of the Taylor & Francis Group, an informa business

British Library Cataloguing in Publication Data
New urbanism : life, work, and space in the New Downtown.
-- (Design and the built environment series)
1. Urban renewal--Europe--History--21st century.
2. Urban renewal--North America--History--21st century.
3. Urban renewal--Europe--Case studies. 4. Urban
renewal--North America--Case studies.
I. Series II. Helbrecht, Ilse. III. Dirksmeier, Peter.
307.3'416'094-dc23

Library of Congress Cataloging-in-Publication Data
New urbanism : life, work, and space in the new downtown / edited by Ilse
Helbrecht and Peter Dirksmeier.
    p. cm.
Includes bibliographical references and index.
ISBN 978-1-4094-3135-0 (hbk.)
1. Urban renewal. 2. City planning. 3. Central business
districts. 4. City and town life. 5. Sociology, Urban. 6. Urbanization.
I. Helbrecht, Ilse. II. Dirksmeier, Peter.
HT170.N49 2011
307.3'416--dc23

2011049399

ISBN 13: 978-1-138-27137-1 (pbk)
ISBN 13: 978-1-4094-3135-0 (hbk)

Contents

List of Figures

List of Tables

Acknowledgements

Editing this book has been a joyful experience because we had the chance to meet and cooperate with wonderful colleagues – some of them we had known before, some of them we got the chance to befriend during the process of producing this book. Therefore we would sincerely like to thank our various companions on this way of collecting and discussing the manuscript. The anthology *New Urbanism: Life, Work, and Space in the New Downtown* points at the challenging task of planning full-scale neighbourhoods on former urban waste lands, harbour areas or deindustrialised spaces. Most of the chapters of this anthology date from a convention called 'Planning Urbanity: Work, Life, Space in the New Downtown' held in Hamburg's HafenCity in March 2008. The book project and the launching conference would not have been possible without the generous financial support of the HafenCity GmbH in Hamburg. Especially Jürgen Bruns-Berentelg, CEO of HafenCity, inspired our reflections on new urbanity and urban renaissance in Hamburg and, thus, we would like to thank him most sincerely for facilitating our ideas on new urbanism in urban renaissance projects.

We also highly appreciated the help of Rebecca Kennison and Sebastian Schlüter in the more mundane processes of putting the collection together. Discussions with both of them have been very helpful to shape the project and make it happen. Thanks also to Valerie Rose and her team at Ashgate Publishing for splendid cooperation. And last but not least we would like to thank all the authors: without them, there simply would be no book at all. We hope that the papers presented in this volume will help to sensitise urban planners and architects as well as urban geographers, urban sociologists and urban anthropologists for the immense task of *planning urbanity* in the New Downtown.

Ilse Helbrecht and Peter Dirksmeier,
Berlin

Chapter 1

New Downtowns:
A New Form of Centrality and Urbanity in
World Society

Ilse Helbrecht and Peter Dirksmeier

Two questions provided the occasion for this volume: which forms of urbanity can be identified for the cities of the 21st century? And can these new forms of urbanity be planned? With these questions, this volume aims straight at the heart of urban research. Urbanity is a classical term which has inspired the imagination and conceptual vitality of the field ever since the scholarly study of cities began, whether in urban sociology, geography, economy or ethnography. All the great thinkers on cities in the 20th century, whether Robert Park (1915), Walter Benjamin (1928/2001), Jane Jacobs (1961/1992), Lewis Mumford (1961) or Georg Simmel (1903), have asked themselves the question: what differentiates urban life from rural situations? And what exactly is urbanity?

The debate on urbanity reached an odd stasis at the end of the 20th century. While, on the one hand, no new theoretically inspired definition of urbanity was being discussed (Dirksmeier 2009), relatively innovative urban developments in local city building projects could be observed at the same time. New types of cities and neighbourhoods arose. However, they were not – or have not yet – been included in a new terminology of urbanity. That is the state of the question from which the present volume proceeds. We are looking for a new type of urbanity in the 21st century. To do this, we will use particular sectional developments in cities as a lens through which to conceptualise and exemplify the new urbanity.

Our thesis is thus that at the beginning of the 21st century, a new type of urbanity presented itself. And this new form of urbanity can, to some degree, be planned. Its contours can be especially clearly seen in a particular, very recent type of city centre, the new downtown. Thus, we'll develop the idea of the new downtown in order to outline a few basic characteristics of the new urbanity of the 21st century. The phenomenon of the new downtown – such is our thesis – is a completely new category of urban area in the globalised world. It is distinguished by at least four characteristics. Firstly, it can be planned to some degree and thus owes its development in large part to deliberate planning and design. Secondly, it is defined by designs which aim to stand out from the traditional geographic imagination of the old downtown, whether in Europe, Asia or America. Thirdly, the new downtown intends to distinguish itself as a representative of a new performative urbanity, and to move existing forms of urbanity in the old downtown forward. And that brings us to the fourth decisive force behind the new urbanity of the new downtown. It is based on a new form of centralisation. This emerging centrality of the new downtown properly belongs to the global conditions of contemporary society. Contemporary society must be described more precisely than as just a world society. It is also the global character of economics, politics and lifestyles, which, with the new downtowns, creates a unique place in the cities for a new form of centrality. A world society opens up a new framework of reference for processes of centre formation which goes beyond the concepts of globalisation and the global city (Sassen 1991). From the perspective of the world society, a new centrality is not just about the control capacity of world markets through transformation of the centre, as Saskia Sassen has aptly established. The world society on the horizon also means a new centrality and urbanity, which can expand economically fixed perspectives and also proceed on the basis of communication of social systems across national borders, from the public realm to world literature to the lifestyle of global milieus.

In this volume, we and the other authors investigate this new quality of inner-urban development with regard to the conditions of a world society. In the following introduction, we attempt to outline a few basic features of this new development. What characterises the new downtown as a new type of urban space? How does it arise? How does it work? And what social consequences and conflicts are inherent to it? In the individual chapters that follow, specific aspects of this new development in city centres will be looked at in more depth.

Why New Downtowns?

The city centre is no longer necessarily in high demand. Instead, the value of the city centre as such can only be proved within a specific social structure. In the ages of feudalism and industrialisation, city centres were traditionally the place where land had the highest price and where the most buildings were constructed (Alonso 1983). This high demand and the social significance of the city that came with it both increased and decreased in recent years. In the wake of globalisation and the metropolisation of urban development, middle classes and elites rediscovered the desirability of proximity to the city centre (Butler and Robson 2003; Ley 1991, 1997; Williams and Smith 2007). This led to an urban renaissance which initially applied functionally and spatially only to the traditional city centres and the adjoining neighbourhoods. Since the 1960s, the mass phenomenon of gentrification (Glass 1964) and the return of the city centres (Helbrecht 1996; Lees et al. 2008; Ley 1980) has been under discussion, initially as a niche phenomenon, and then by the 1990s at the latest as the mainstream of urban development in both Europe and North America as well as Australia. On the other hand, the status and functionality of the centres have again changed remarkably in the wake of globalisation and the development of the world society. Through processes of disembedding

(Giddens 1996: 33), globalisation leads to a loss in the significance of old centres due to decontextualisation.

The phenomenon of the new downtowns is situated precisely at the intersection of two movements: the increased significance of old centres due to gentrification with the formation of a new centralism, and a changed urbanity under the conditions of a globalised contemporary society or world society.

The current social relations can be most accurately described by a type of society which has evolved from the service society, but relinquishes the demand for a relational social unity and no longer has recourse to an – unrealistic – precondition of homogeneity and internal connectivity as essential constituents for a society. The term world society actually describes this type of society (Burton 1972; Heintz 1982; Luhmann 1975/1991; Stichweh 2000). Describing contemporary society as a world society does not negate the existence of nation states; it does, however, serve to emphasise that nation states are structurally identical in a variety of unexpected dimensions and that they are evolving along similar lines (Meyer et al. 1997: 145). State boundaries remain important but are no longer a determining factor: 'State boundaries are significant, but they are just one type of boundary which affects the behaviour of world society' (Burton 1972: 20). Particularly with regard to future developments in the 21st century as well as questions concerning a future urbanity, it is necessary to think about the perspective of the world society as a premise for the development of urban spaces and centres. The contemporary type of society is no longer organised as a differentiation between the centre and the periphery in the same way that it is not structured as an industrial or service society. Modern society can be considered much more strongly in terms of a world society and as the best system of organisation for all structures and processes of social systems currently existing at a global level (Stichweh 2000: 11–12). Without the globalisation of cultural codes, material products, financial services and political conflicts, it is no longer possible to understand local developments. The term globalisation on its own, however, is not enough to describe the quality of this type of society. It refers more to the situation within which expansion occurs or to the de-localisation of heretofore limited phenomenon without registering the new system of an all-encompassing quality which the phenomenon of globalisation uses for its own creation (Stichweh 2000: 14).

World society possesses this all-inclusive quality which serves as the most extensive system of inter-connecting communication. But this does not necessarily mean that a sort of 'one world' will now emerge. The hallmark of a world society is characterised more so by a lack of homogeneity. That is why a global framework of reference is relevant for nearly all differentiations, structural formations and processes, in spite of and particularly because of all noticeable peculiarities, all absence of synchronicity throughout historical development as well as all the spatial differences which appear in world society and which are caused by it (Bahrenberg and Kuhm 1999: 194). With the collapse of society into a single emerging system of communication, it is important to focus our attention on traditional urban society.

Urban spaces still exist and they are growing. Nonetheless, there are no longer any spatial barriers between more or less autonomous, co-existing societal segments. Centrality no longer automatically derives from the social order. Hence, urban society cannot simply be distinguished from something in its environment. But world society continues to establish urban spaces in local–global communication processes at an intense level.

The two-pronged demand for a) social goods as well as b) the economic product 'city centre' has, in specific cities and specific situations, produced a run on the development of city centres and their surrounding neighbourhoods. For those cities which profit from the special conditions of the new centrality of a world society, and also, due to their location near the sea or on a lake or river, have the advantage of a large industrial district and old industrial waterfront districts near the city centre, there are special opportunities. Such cities carve out a niche for the new centrality of the globalised world society near their old downtown. Thus, large-scale urban projects try, through waterfront redevelopment for example, to produce local centrality for global dimensions. In such cases as New York Battery Park, the London Docklands, the Barcelona Strand, in Buenos Aires or the Hamburger HafenCity, local resources can be used to creatively meet the new demands placed on the city centre by means of expansion. Young areas arise in direct proximity to the old city centre: the new downtowns. They have a certain fascination for businesses, planners, residents and researchers due to their unique spatial and historical situation. Very near, sometimes just a stone's throw from the old city centre, an extraordinary process takes place on the remains of the industrial society and in the former warehouse districts and ports: inner-city expansion is taking place in inner-city locations using existing spaces. This is urban development under growth conditions. The proximity to the old centre and the concomitant significant spread from it characterise the locational strategy of the new downtowns. This is a paradoxical gesture, as is, according to Lefebvre (2003), typical for urban processes.

Table 1.1   Characteristics of the old and new downtown

Old downtown	New downtown
Old centrality	New centrality
Historical growth	Strategic planning
Construction layers from diverse eras	Representation of the planning and architectural style of one decade
Functional homogeneity from the late 20th century	Mixed function through planning
Service society	World society
Embedded in local tradition and regional networks	Global networks as a strategic impulse, symbolic simulation of local tradition
Visiting card and object of identification for the whole city (tourism)	Object of identification in the system of global cities (elite identities and elite discourse)
Small-scale structures	Large-scale planning
Loss of residential function	Cultivation of residential function
Fragmented building structures	Ordered building structures
Local and regional agents	Global capital and global agents
Traditional gender roles	New gender roles
Modern public space	Postmodern public space

Source: Own conception.

The comparison on the basis of these descriptive differences shows that the category of the new downtown can only be understood in terms of its new social functions and global embedding.

In the Chicago School, especially for Ernest W. Burgess (1925/1967) the city centre was more than just the spatial centre of the city. Instead, the city centre had contained all functions and social classes of city's population at an earlier point in the development of the city. It was the functional centre of the city. Only after the process of both centrifugal and centripetal expansion did the modern metropolis with its differentiated structures form. The city centre is thus, for the Chicago School, the cradle of urban development. It functions like an incubator: it is a dynamic centre and central motor for all the other processes. In the course of the 20th century, however, the old downtown in Europe, Australia and the United States and Canada initially lost significance and function for the city as a whole. The decline began with the loss of the residential function, which took on an increasingly one-sided usage structure and with that impoverishment due to the dominance of retail usage. Once retail had reconstituted itself in chain stores, the inner city seemed to have lost its special character.

Even small businesses moved out to the suburbs. The decline of the old downtown thus seemed certain. Resistance on the part of many old downtowns to regain significance by building inner-city shopping centres to emulate the retailer's suburban malls is, especially in Europe, a desperate attempt to copy a one-time suburban success in the old city centres. Most architects, geographers, urban sociologists and planners are aware that this is not an adequate way to transition the old city centre into a new epoch. The 'malling' of downtown would simply be a poor copy of the suburbs or surrounding towns instead of an original inner-city development (see Garreau 1991; Goss 1993; Popp 2004). The real significance of contemporary development of inner-city locations can be seen more precisely in the new downtowns. In this historical watershed, in which city centres have gone through a cycle of crowding, poverty and through the rediscovery of urban renaissance, it is an exciting question for urban research to compare the significance, formation and structure of the old and new downtowns.

What is decisive here is to understand the changes in function and significance in the centres which have occurred since the beginning of the 20th century. Even if many of the prerequisites for urban research according to the Chicago School no longer apply, one basic assumption remains an interesting point of departure for urban geographical research: how can city development be construed from the perspective of the centre? And what actually defines the centre of a city? Our suspicion for the 21st century is that urban spaces today must still, but not exclusively, be understood from the perspective of the centre because the urban principle itself is about forming centres and being central. The city centre can be an extraordinary place within the city which encourages discussion regarding its position and composition. It is not an uncontroversial place, because it is always a creative place, a 'place for creation' (Lefebvre 2003: 28). What makes urban places special in contrast to non-urban places is the collection of diversity.

> What does the city create? Nothing. It centralizes creation. And yet it creates everything … The city creates a situation, the urban situation, where different things occur one after another and do not exist separately but according to their differences. (Lefebvre 2003: 117)

The core of the urban situation is thus the possibility of centrality by means of the connection and relation between diverse contents in one place. However, in the globalised contemporary society there is no longer a privileged place for the centre. Centrality itself has become mobile:

> The essential aspect of the urban phenomenon is centrality, but a centrality that is understood in conjunction with the dialectical movement that creates or destroys it. The fact that any point can become central is the meaning of urban space-time. (Lefebvre 2003: 116)

The form of the city of the future will thus, especially in times of urban renewal in which the old downtown is restructured and re-invented, be derived from the form, content and shape of the new places of centrality. In this constant movement of centrality and the contradictory calls for a new centrality while the centre's functions are being dissipated, the development of the new type of urban space, the new downtown, can be oriented. Before we go into the details of this new driving force and the structures of the new downtowns with the authors of this volume, the concept of the new downtown itself must be introduced. What is a new downtown? To what extent is it different from the old city centre? How is its urbanity articulated? And which features can be planned?

Copenhagen and Istanbul:
Two New Downtowns as Visions of Urban Development

There are many examples of new centrality projects around the world, in which new downtowns have deliberately been made. As a point of departure for our theoretical considerations of the phenomenology, functionality and centrality of new downtowns, we would like to begin with two exemplary projects of the new urban renaissance. These projects should familiarise us conceptually as well as practically and visually with the phenomena of the new downtown. The examples are the developing area Örestad in Copenhagen and the Kartal Business District at the interface of Asia and Europe in Istanbul.

In Copenhagen, a revitalisation project which can be described as a new downtown extends over 210 hectares between the old city centre and the international airport. Under the supervision of the Örestad Development Corporation, urban expansion is taking place through four sub-projects and is pursuing the ambitious goal of placing Copenhagen on the international map of global cities. The project began in 1996, and actualisation is planned for 2017. Örestad is thus a new downtown which is in transition from both one defined urban structure to the next as well as from city and country. Spectacular architecture, including the work of British architect Sir Norman Foster (McNeill 2005), is interspersed with open public spaces and parks. The area is accessible by metro. Several universities have already moved into their new buildings, and a new concert hall with 1,800 seats has also been built here, while the broadcasting centre of the Danish Broadcasting Corporation (Evans 2009: 1027–8) is one of the largest media buildings in the world. With this development concept, the new downtown Örestad attracts a young, international public, as sociological research has confirmed. Every third resident is under the age of 30, and a further 28% are between 30 and 39. These young residents of Örestad particularly value the impressive architecture and the proximity to both the city centre of Copenhagen and the beaches and recreational areas on the outskirts of the new downtown.

**Table 1.2   Overview of the stages of development in Kopenhagen/
                Örestad, March 2008**

	Surface area in m²	Percentage
Office space	64,700	10.5
Housing space	271,700	44.1
Services	101,900	16.5
Shopping Mall	178,000	28.9
Total	616,300	100

Source: Majoor and Salet (2008: 100).

The southernmost, 50 hectare development area of the project is planned
for 10,000 residents and 20,000 employees. This functional mix ties it to the
neighbouring urban spaces which serve as generators of diversity (Jacobs
1961/1992: 143) and are an important element within the new downtown Örestad
for making Copenhagen an international metropolis (Priebs 1992). In South
Örestad, a total of 1.2 million square metres of floor space is planned and will
consist in equal parts of housing and service spaces. The buildings and public
areas, with a classic hierarchical structure, have an aesthetic dimension, and the
public squares, residences and places of work have a physical dimension. In South
Örestad, particular value is placed on urban change, which, in an architectural
sense, indicates the relationship between external and internal spaces.

Kartal Business District, Istanbul: An Osmanian Alternative

The Kartal Industrial Area Central Business District Plan can be considered one of
the most ambitious urban building projects in west Asia. With it, the city of Istanbul
aims to physically consolidate its growing global influence. The entire area of the
Kartal Business District includes 555 hectares on the Asian side of Istanbul. The
project is a new downtown which offers 120 hectares of floor space for residences
in 15,000 new homes intended for 70,000 people. The architecture, planned by
Zaha Hadid in a wave-shaped horizontal structure on a grid format, speaks a
spectacular international language. The entire area stretches from the Marmara
Sea in the south to the E5 motorway in the north. The realisation of the project is
planned from 2009 to 2026. The Kartal Business District is intended as a genuine
second city centre, connected to the old city centre of Istanbul on the European
side via motorway, metro and ferries (Hadid 2008). The Kartal new downtown
aims at a redistribution of the economic power of Istanbul, which until now has
been concentrated on the European side of the city, while the Asian side has been
characterised by suburban bedroom districts. The city administration is planning
the Kartal Business District as a holistic neighbourhood with a central business
district, high-end living space, concert halls, theatres, museums, recreational
facilities and a leisure area in a former stone quarry which is being made into a

lake with a marina and first-class shops and restaurants. The aspiration connected with the project aims at nothing less than the establishment of a poly-central city structure. The goal is, however, still a 24-hour city, i.e. urban life without time limits (Melbin 1978). With that, Istanbul clearly articulates its ambition of building a new downtown as an already gentrified counterpoint to the old, no less gentrified, city centre (Potuoğlu-Cook 2006). The new downtown consists of a north-to-south axis, at both ends of which a high-rise financial district is located. Around these financial centres are recreational areas of various qualities. In the north there is the waterfront around the lake in the former stone quarry. The banking district in the south is hemmed in by the marina with its shops and restaurants. The Kartal Business District is a holistically planned new downtown, which is to be equal to the old city centre on the other side of the Marmara Sea in its economic power, living and shopping facilities, as well as its recreational and leisure areas (Hadid 2008).

Preliminary Conclusions

As these two examples show, the new downtown is obviously a global phenomenon in which local or regional particularities, apart from architectural styles or local limitations, cannot be discerned. The new downtown is a type of global urban development which, one might suspect, consists of its own form of urbanity. The basic problems of typology arise out of the unclear connections which the types themselves form in and among each other. There is a problem with sorting things into categories. Not only do the things exist in unclear relations to each other, but so also do the categories which are actually supposed to order the things (Erdmann 1894: 20). We would thus first define the category of the new downtown more precisely according to its size, functional mix, connection to world society, and planning, and then move on to the question of urbanity in a second step.

A New Urbanity and New Centrality for the New Downtown

What is new about new downtowns? What sort of new centrality and urbanity are being realized? And how exactly is it different to the well-known work of Georg Simmel, Louis Wirth and Jane Jacobs, all three of whom saw urbanity as the consequence of the city's variety and diversity in a concentrated heterogeneous space? The new downtowns have some new attributes in their waterfronts, parks, festival markets, multi-purpose arenas and cultural centres (Ford 2003). But are they actually more functional, lively, exciting and truly new in comparison to the central business districts and cities of the 19th and 20th centuries? There are four arguments which speak for a fundamental difference between the category 'new downtown' and the city centre as perceived through the eyes of the leading names of urban research such as Simmel (1903), Park (1915), Wirth (1938), and Jacobs (1961/1992). These are:

1. world society – the new quality of centrality;
2. stage aesthetics – the changed role of architecture;
3. flagship policies – urban planning as globally oriented sectional policy;
4. performative urbanity – fleeting events.

New Centrality

The classics of urban research were working on the basis of social forms which allowed them to assume a significant centrality of the city as such. Whether in antiquity, the middle ages, or the industrial period, the city was a political, religious and economic centre of power. This spatial order was supported by the historically diverse social orders in various ways. In each case, the issue was social order organised in tension between the centre and the periphery. Social centre–periphery relations were articulated in urban spaces as mirrors and moulders. Thus the middle ages, with its stratified society, was organised hierarchically from the court outwards. The medieval state included concentration of the population and of resources in the main cities, which also had a hierarchical urban society. This concentration was also carried forward in a second step of differential regulation. The actual inequality between the centre and the periphery in the availability of resources was normatively codified, since different rules and regulations applied to the centre and the periphery respectively. In the 18th century, for example, Moscow was subject to different legislation than the rest of Russia. In 19th-century Europe, there are higher standards of qualification for various urban professions such as the medical profession (Stichweh 2006: 494–5).

As it arose out of feudal society, industrial society ceased codifying the differentiation of the centre from the periphery, but its structure led to the creation of economic centres, which were centres of production but not necessarily of politics or culture. This is a qualitative difference to stratified feudal society. Industrial society had recourse to the masses in both objective and human terms. The human masses of industrial society first manifested their political power when physically assembled in a given place. Peter Sloterdijk even goes so far as to claim that in this case, the masses themselves form spaces for the masses and thereby appear as an active subject (Sloterdijk 2000: 17–19). Such a mass of people gave birth to a new quality of social life. The appearance of an assembled mass of people in front of itself as well as for itself was one of the key moments of industrial society (Sloterdijk 2000: 16). Industrial society also created masses of objects as a good, whether as raw materials, work or the market. These masses of things required centrality for their organisation, as Lötsch writes: 'it is essential for the regional system of market areas to have a center' (1938: 77). Centrality and industrial society constitute a unity which offered good conditions for satisfying the needs of industrial society by means of trade and market centres (Reulecke 1985: 43).

The category of the service society as a subsequent societal formation loses this natural quality of centrality as was characteristic of the industrial society.

Lefebvre's theory that it is formally possible for centrality in a city to arise anywhere is also supported by urban researchers such as Saskia Sassen (Sassen 1991). This relates to the fact that small-scale agglomeration sites can be equated with every agglomeration factor which is effective on large-scale inter-city levels. With regard to both dimensional scales, similar mention is made of the role of network relations in the information economy. As a slogan of the information society, tertiarisation is therefore more dependent on functioning networks and less so on centrality. Thus, 'office nodes' emerge at airports, for example, or as highly-specialised office locations which differ greatly in terms of sector, size, type and number of employees. These nodes can occur spatially almost anywhere and are therefore, from a territorial perspective, subject to trends which can both create provisionally functional centres just as quickly as they watch them disappear (Friedrichs and Kiehl 1985). The exchange of goods and services generates their centrality which, in turn, is measured according to the intensity of this exchange (Barton 1978: 37). The compression of time and space, or equally the end of geography, are the key words of this society. The result has been a qualitatively new form of centrality which has acquired its trademark as a global city (Abu-Lughod 1999).

In this, centrality is currently and prospectively more like a commodity that must be produced and reclaimed. The social order no longer supplies the city with its own unique centrality (Garreau 1991). The edge cities in North America are an expression of spatial relations within the world society. Political mechanisms of organisation no longer determine spatial organisation; in this case, it is more those factors pertaining to the market economy and lifestyle which define the form of the city and centrality. Edge cities are therefore becoming decentralised in an ongoing process (Ding and Bingham 2000). Centrality thus becomes a marketable good, a rare commodity whose value is measured in terms of a spatial advantage which is capable of generating itself, as explained by the French sociologist Pierre Bourdieu (1991). According to Bourdieu, social and spatial positions can be seen in terms of an equivalent relationship. In other words, the most desired spatial positions, such as residential areas or office locations, can only be inhabited by the occupants of equivalent social positions. It is thus that these places generate spatial advantages in the form of kudos which arises from the interpretations of onlookers concerning the spatial proximity of positions to the most desired and valuable locations as well as concerning spatial proximity itself (Dirksmeier 2007).

The French philosopher Henri Lefebvre put forward a similar argument, but with a slightly different emphasis. According to Lefebvre, urban spaces enable centralisation per se. Urban constructs are therefore actually merely forms of agglomeration which stand in a dialectical relationship to that which is centralised on location: 'The urban is, therefore, pure form: a place of encounter, assembly, simultaneity. This form has no specific content, but is a centre of attraction and life' (Lefebvre 2003: 118). These centres can dissipate themselves, can dissolve themselves, can fall into decay or recreate themselves. 'This is why urban space is so fascinating: centrality is always possible' (Lefebvre 2003: 130).

That is also why 'the right to the city', as expounded by Henri Lefebvre, is an aspiration not to be excluded from centrality and its dynamic (Lefebvre 2003: 150). According to Lefebvre, the contradictory nature of all things urban has its roots in the fact that new contents are constantly being accumulated in urban spaces and creating centrality. At the same time, however, the centres are disseminating and dispersing that which is being centralised. Thus, the right to be a centre and, with that, a place of gathering onto as well as giving from itself is currently being socially tested, as is the location and content of centrality as such.

Centrality is therefore something that is in flux. The topographical locations of centres shift within a city as well as between cities. And the contents that are accumulated within a centre change. Concentration and dispersion take place simultaneously. There is a social struggle taking place as a result of both power and class interests regarding exactly what and which place is allowed to be considered central. The formation of a centre, the ultimate urban form, is thus a political process as well as a question of social, cultural and economic power.

New downtowns represent the claim of spatial advantages in world society with regard to the production of centrality. It is the new downtowns that create centrality, thereby generating spatial profits which, in turn, are competed for in world society. The legitimate appropriation of locations in the new downtowns secures symbolic capital, which a global elite in the form of a class of gentry are competing to share. This global elite use the spatial advantage of the new downtown as part of its constructed image as both 'global' and distinctive (Rofe 2003). In this, new downtowns have turned the tables on the cause–effect relationship behind the classic urban centrality: centrality is no longer a product of social order, but more the result of an artificial production of spatial advantages arising from the creation of valuable locations within world society.

Architecture as Stage

This shift in meaning has attributed a new role to architecture, to aesthetic urban design. This comprises our second argument: an increase in the significance of architecture and the physical settings of centrality. The design of buildings, the names of architects and the symbolic capital that they represent were historically always essential factors for the process of generating centrality along with its attending spatial profits. Formerly, however, it was largely the segmented or functional centre–periphery differentiation which predetermined the location of a centre. Owing to the conditions of the world society, classic location attributes are losing their meaning. With the help of his time-space-compression theory, David Harvey (1990) indicated that due to the collapse of spatial borders and the fading importance of distance, more attention is being focussed on that which actually occupies the spaces. Location itself is fading into insignificance, while the meaning of architecture and urban space is asserting itself once more as the stage for a new centrality. Given that a quasi-natural city centre no longer simply arises from the attribute of location, but that potentially all urban segments struggle for supremacy, the aesthetic architectural representation and conquest of centrality is becoming more important. The symbolism of urban architecture has most decidedly been designated the task of urban centrality design, no longer simply as a representation of the centre, but more importantly as part of the process of creating the centre itself.

What is more, the design of buildings, streets and squares should allow for performances and in so doing should draw its strongest trump card, namely the presence of others. With reference to Erving Goffman, performance describes a public demonstration of artistic or social acts in the form of a unique and short-lived event. Performance is thus 'the embodied enactment of cultural forces' (Carlson 2004: 212). From a statistical point of view, architecture influences the success of a city in two ways. Architecturally attractive cities grow faster than unattractive cities and cities which can be quickly traversed also grow faster (Glaeser et al. 2001). New downtowns are supposed to possess a flagship architecture which is attractive, but still permits fast transit; they are also supposed to allow for performances and the appreciation thereof as well as guarantee spatial advantage. Franz Dröge and Michael Müller coined a special term for this: the exhibited city (Müller and Dröge 2005). In so doing, they try to express the idea that, similar to exhibition politics in museums, cities increasingly see themselves as exhibits and display themselves as such. Thus, architecture becomes the discipline of conservation and the city planner functions as a curator.

Global Flagship Policies

This leads to our third argument: urban politics as both flagship and district politics. The political planning concerning the design of a new city centre, a new downtown, is as much a dream as it is a nightmare, because the development of a new downtown is inconceivable without a clear and strong public impulse from urban politics. New downtowns are defined in part by the fact that they owe their creation to the deliberately planned design and far-reaching political intervention of the city. They are large-scale projects in terms of finance and urban planning. In contrast to the time of Fordism, however, these projects are not arising on the drawing boards of the city council's urban planner, but are being developed and conducted as large-scale projects in close networks of cooperation between public and private agents. It is a question of *urban management* as opposed to *urban planning*. The public hand has shown itself in the cases of cities such as Hamburg or Barcelona as increasingly economically professional in its dealings with its most important asset: planning power to create new locations and patterns of use for a new centrality. With that, the emergence of new downtowns in many cities is a perfect example for new urban governance.

At the same time, the public hand is also taking a significant risk with the design of such new locations and the artificial creation of a new centre. From a financial as well as urban-design point of view, the completely new layout of a downtown area represents dangers of miscalculation, misplanning and misjudgement on all fronts: financial, economic, cultural, ecological, and in terms of urban space. Nonetheless, within the context of the urban renaissance of the last few years, more and more cities are prepared to take the risk of a completely new development of a district. With regard to this, the most recent standards and character of urban renaissance projects have been visibly internally differentiated as well as enlarged on in terms of the extent of urban planning. In the HafenCity of Hamburg, an expansion of the inner-city area is currently been ventured in the new downtown with a view to enlarging the existing inner city by up to 40%. Should one analyse the motives for this quantitative and qualitative increase in dimensions with respect to the politically planned design of new downtowns, then one could identify three leading driving forces common to all projects:

Exemplarity – pars pro toto   The new downtown tries to be what John Friedmann calls an 'open city' (Friedmann 2002) despite its territorial limits: the new downtown wants to reduce excluding structures, to strengthen the residents as citizens by means of governance projects and with that cater for the basic needs of the residents in a way that is socially compatible. As with the 'open city', the new downtown is an experimental project (Friedmann 2002: 238) which allows the city district to become a unique exhibition space. Just as a museum needs unusual exhibitions in order to re-invent itself, the old city re-invents itself in its new downtown district. Within the context of international urban competition, many metropoles create a name for themselves on the international stage by means of the strategic design and marketing of individual city districts. The district becomes the *pars pro toto*. It represents the spirit of optimism in the metropolis, its reorientation and economic as well socio-cultural potency. This is what makes new downtowns ideal-typical flagships for metropolis development by means of *pars pro toto*. The international visibility of a city like Barcelona, Buenos Aires or London is heightened by the representative development and marketing of a singled-out spatial section. Should this, however, not simply be a question of a traditional waterfront development by means of emphasising living, services and pleasure, but that the design of a new downtown is truly daring in the sense of a completely new typology for urban planning, then the effectiveness of this *pars pro toto* strategy is even greater.

Monumentalisation: The unique exhibition space of the city   Since the 1990s, there has been an important discussion concerning the extent to which the tourist's visual gaze (Urry 1990) influences the perception of a city through the eyes of its inhabitants and its visitors as through the eyes of its businesses and employees. Planning's attempt to react to this growing need is leading to a wave of monumentalisation (Groys 2000). Monuments are being created which, similar to a visit to the Roman Colosseum or Manhattan's Central Park, aspire to historical grandeur and originality of design. The tourist's visual appetite and its consequent craving for monumentalisation is characterised by an ambivalent need: 'on the one hand, we constantly seek that in our city which most embodies the dream of immortality, while at the same time, we wish to reconstruct, improve and even demolish it' (Müller and Dröge 2005: 191). Attempts are being made in the new downtowns to create architectural and urban-like monuments which will capture the gaze of even the most demanding globetrotter. A good example for this can be seen in the Elbe Philharmonic of the HafenCity in Hamburg.

Regeneration: The phoenix rises from the ashes   Not least and quite obviously, the high level of commitment from politicians and planners alike in the design of new downtowns can be traced back to an insatiable drive to act in the interest of urban regeneration. Old industrial areas, dilapidated harbour areas etc. which render the task of structural development within a city tangible, offer an opportunity for public care and attention in these zones. At the same time, their advantageous location offers a unique opportunity to undertake extensive inner-city urban development. Large-scale projects aim at attracting global attention under the multifarious auspices of planning, because the new downtown creates a cluster which aspires to be the centre of society. New downtowns are becoming the stages of urbanites (Mumford 1996), on which they can perform unnoticed and with optimum kudos and speed. The issue involved here is not the provision of spatial opportunities to perform, which were surely already well-known in the city, but more so the framework of these performances. Architecture, planning and district politics serve as an aesthetically and functionally optimised framework for performances. Should these criteria coincide within a district, then centrality arises within the world society.

Performative Urbanity

With that, we come to our fourth argument: the newly evolving quality of a performative urbanity. For it is increasingly the fleeting and momentary performances which are perceived as urbanity and which make it onto the stage of the new centres of the world society.

Cities bring people and things together in a multiple number of ways. The presence of the unknown is its greatest fascination (Meyer 1951). In a more recent essay on the street as a 'stage' for performances, Richard Schechner enquires about the relationship between the planning and political authorities of a city and the various people who start to affectively occupy and perform in public streets, squares and buildings (Schechner 1993). As an expert in theatre studies, Schechner sees a natural shift of emphasis in this correlation from the side of planning to the side of performance. In his eyes, streets are turning into theatrical stages on which people are vying for the attention of others, more or less regardless of the existing plans for land-use or development, of the property rights, site rules or regulative architecture. That fleeting moment of capturing the attention of others is a rare commodity which is being competed for on the various stages of the city and with various motives.

According to cultural geographer Nigel Thrift, misanthropy as an anthropological constant is a driving force behind this phenomenon (2005). Life in the city continuously provokes feelings and emotions which can be interpreted as subliminal misanthropy. Humans cannot help themselves; from time to time, they have to disparage or emotionally denunciate others. At the same time, they need the space in which they can vent these feelings. And so, urbanity feeds off the fleetingness of attention, which in itself is merely a misanthropic performance manifested in the disregard for beggars, the cursing of careless motorists and cyclists among others. Many of the key moments in the city are the result of coincidences, of completely dysfunctional juxtapositions and these, in turn, facilitate the threat of misanthropy which weaves its way through the modern city. Urbanity is just as much a mistrust and an avoidance of others as it is the desire to capture their attention, for example, by means of extravagance (Hofmann 2004) and fancy fashion (Esposito 2004). What is more, sensory inputs present themselves in steady and broad streams (Helbrecht 2004). People in the city focus their attention on the detail of the performance of others. The performance taking place on Schechner's street stage is aestheticised, while at the same time tinged with emotion. This emotionality results from the simultaneous mixture of moment, fleetingness and the construction of meaning. It is no longer what Simmel refers to as the urbanite's ability to dull his senses, i.e. his ability to perceive selectively within the structural and social density, which is significant, but more his sensibility for performances on the city's stages which is required. It is not the use of complacency as a means of limiting perception, but the sensuous recording of performance in all its facets in any one particular moment which distinguishes the urbanite. The urbanite treads the streets or stages of the city and expects individual original performances. The urbanity which results from this expectation is thereby just as transient as are the performances which originally constituted it themselves.

It is not the enduring social qualities of a city as elaborated on by the classics, headed by Jane Jacobs, which are decisive for urbanity, but more so those social qualities of a transient nature. This form of urbanity is a new one because of its performative quality. It produces itself in the fleeting performances which range from misanthropy to philanthropy. Performative urbanity creates itself from the urbanite's attention for transiency on the stages of the city, the rapid transition which demands attention and which can be, but must not necessarily be, planned. The new downtown offers the framework for these misanthropic or philanthropic performances. Their urbanity is performative. The performative urbanity of the new downtown is the keen attention of the urbanite for the fleetingness of the observed performances.

When all this is considered, the new downtowns as urban development projects of the new *urban renaissance* are truly different from the old inner city. The difference can be attributed to the four points as outlined above.

1. Centrality is no longer the result of a social order, but is both produced and marketed in the new downtowns.
2. Architecture has been assigned the role of generating spatial advantages, while at the same time providing space for misanthropic and philanthropic performances.
3. Urban politics can be more accurately described in relation to projects as district politics.
4. And the urbanity of the new downtown is performative. It is the keen attention for the transiency of performances.

Statistic surveys on life in Great Britain's inner city reveal that it is specifically the simplification of everyday life, the easy accessibility of the norm and not the exception (such as cultural events), which contribute to the appeal of the inner city (Tallon and Bromley 2004). New downtowns should not only offer a balance between the usual and the unusual, but also spatial advantages and 'stages' if they wish to emphasise their most significant distinctive feature: the economic valorisation of centrality.

Can Urbanity be Planned?

The vision of being able to plan a new urbanity harks back in a broad sense to an old dream of mankind. It is the Faustian wish of humans to fully create their own environment in such a way that they can live in it happily and fulfilled forever more. To what extent can project developers and investors, planners, home builders and retailers foresee and deliberately design the complex form of a functioning new inner city? To what extent can the new urbanity of the new downtown be professionally produced?

In this volume, we want to examine both the question of planability concerning a new urbanity and its presumed, observable and currently embryonically developed form. We use different points of view in order to do this. We enquire about the economic impetus of a new centrality (Chapters 5, 6, 7 and 8) as well as the motives of state intervention with regard to planning the new centre (Chapters 2 and 4). The emergence of new urban lifestyles as target groups (Chapters 9 and 10) and the changing function and form of public spaces are central (Chapters 3 and 11). In this regard, the HafenCity of Hamburg serves us and many authors of this volume frequently, but not exclusively, as a focus of reflection. Hamburg's HafenCity is at present the largest of the inner-city projects in urban renaissance throughout Europe. In literature covering urban research to date, however, it has hardly been comparatively considered with the London Docklands or New York. It is therefore used here as a burning glass through which some authors try to see into the future of the new downtown. But developments in the United States (Chapter 2), Canada (Chapter 4) and the Netherlands (Chapter 5) are also integrated and reflected on. In conclusion, the question of diversity, the central dimension in classic urbanity with regard to the social role of new downtowns, is discussed (Chapter 12). It is our aim, as well as that of the authors of this volume, to cast a critical eye on an ambitious endeavour worldwide in many new downtowns: namely, to anticipate the unpredictable form of the future of human cohabitation in the design and implementation of a new downtown.

References

Abu-Lughod, J. 1999. *New York, Chicago, Los Angeles: America's Global Cities*. Minneapolis: The University of Minnesota Press.

Alonso, W. 1983. A Theory of the Urban Land Market, in *Readings in Urban Analysis: Perspectives on Urban Form and Structure*, edited by R.W. Lake. New Brunswick: Transaction Publisher, 1–10.

Bahrenberg, G. and Kuhm, K. 1999. Weltgesellschaft und Region – eine systemtheoretische Perspektive. *Geographische Zeitschrift* 87(4): 193–209.

Barton, B. 1978. The Creation of Centrality. *Annals of the Association of American Geographers* 68(1): 34–44.

Benjamin, W. 1928/2001. *Einbahnstraße*. Frankfurt am Main: Suhrkamp Verlag.

Bourdieu, P. 1991. Physischer, sozialer und angeeigneter physischer Raum, in *Stadt-Räume*, edited by M. Wentz. Frankfurt am Main and New York: Campus-Verlag, 25–34.

Burgess, E.W. 1925/1967. The Growth of the City: An Introduction to a Research Project, in *The City*, edited by E.W. Burgess, R.D. McKenzie and R.E. Park, Chicago and London: University of Chicago Press, 47–62.

Burton, J.W. 1972. *World Society*. Cambridge: Cambridge University Press.

Butler, T. and Robson, G. 2003. *London Calling. The Middle Classes and the Remaking of Inner London*. Oxford and New York: Berg Publisher.

Carlson, M.A. 2004. *Performance. A Critical Introduction*. New York: Routledge.

Ding, C. and Bingham, R.D. 2000. Beyond Edge Cities. Job Decentralization and Urban Sprawl. *Urban Affairs Review* 35(6): 837–55.

Dirksmeier, P. 2007. Urbaner Raum und Inklusion: Zu einer Paradoxie der Moderne. *Geographische Zeitschrift* 95(4): 199–210.

Dirksmeier, P. 2009. *Urbanität als Habitus. Zur Sozialgeographie städtischen Lebens auf dem Land*. Bielefeld: Transcript Verlag.

Erdmann, B. 1894. Theorie der Typen-Eintheilungen. *Philosophische Monatshefte* 30: 15–49 and 129–58.

Esposito, E. 2004. *Die Verbindlichkeit des Vorübergehenden: Paradoxien der Mode*. Frankfurt am Main: Suhrkamp Verlag.

Evans, G. 2009. Creative Cities, Creative Spaces and Urban Policy. *Urban Studies* 46(5/6): 1003–40.

Ford, L.R. 2003. *America's New Downtowns: Revitalization or Reinvention?* Baltimore: The Johns Hopkins University Press.

Friedmann, J. 2002. City of Fear or Open City? *Journal of the American Planning Association* 68(3): 237–43.

Friedrichs, J. and Kiehl, K. 1985. Ökonomische Phasen der Stadtentwicklung. Eine Anwedung des Fourastié-Modells in einer vergleichenden Studie. *Kölner Zeitschrift für Soziologie und Sozialpsychologie* 37: 96–115.

Garreau, J. 1991. *Edge City. Life on the New Frontier*. New York: Anchor Books.

Giddens, A. 1996. *Konsequenzen der Moderne*. Frankfurt am Main: Suhrkamp Verlag.

Glaeser, E.L., Kolko, J. and Saiz, A. 2001. Consumer City. *Journal of Economic Geography* 1(1): 27–50.

Glass, R. 1964. Introduction. Aspects of Change, in *London: Aspects of Change*, edited by Centre for Urban Studies Report 3. London: MacGibbon and Kee, xiii–xlii.

Goss, J. 1993. The 'Magic of the Mall': An Analysis of Form, Function, and Meaning in the Contemporary Retail Built Environment. *Annals of the Association of American Geographers* 83(1): 18–47.

Groys, B. 2000. Die Stadt auf Durchreise, in *Perspektiven metropolitaner Kultur*, edited by U. Keller. Frankfurt am Main: Suhrkamp Verlag, 60–75.

Hadid, Z. 2008. Kartal Business District, Istanbul. *Stadtbauwelt* 99(180): 54–7.

Harvey, D. 1990. *The Condition of Postmodernity: An Enquiry into the Origins of Cultural Change*. Cambridge: Blackwell.

Heintz, P. 1982. *Die Weltgesellschaft im Spiegel von Ereignissen*. Diessenhofen: Ruegger.

Helbrecht, I. 1996. Die Wiederkehr der Innenstädte. Zur Rolle von Kultur, Kapital und Konsum in der Gentrification. *Geographische Zeitschrift* 84(1): 1–15.

Helbrecht, I. 2004. Bare Geographies in Knowledge Societies – Creative Cities as Text and Piece of Art: Two Eyes, One Vision. *Built Environment* 30: 191–200.

Hofmann, M.L. 2004. Georg Simmel (1858–1918). Theorie der Extravaganz als Kulturtheorie der Moderne, in *Culture Club. Klassiker der Kulturtheorie*, edited by M.L. Hofmann, T.F. Korta and S. Niekisch, Frankfurt am Main: Suhrkamp Verlag, 31–47.

Jacobs, J. 1961/1992. *The Death and Life of Great American Cities*. New York: Random House.

Lees, L., Slater, T. and Wyly, E. 2008. *Gentrification*. London and New York: Routledge.

Lefebvre, H. 2003. *The Urban Revolution*. Minneapolis: The University of Minnesota Press.

Ley, D. 1980. Liberal Ideology and the Postindustrial City. *Annals of the Association of American Geographers* 70(2): 238–58.

Ley, D. 1991. The Inner City, in *Canadian Cities in Transition*, edited by T. Bunting and P. Filion. Oxford: Oxford University Press, 313–48.

Ley, D. 1997. *The New Middle Class and the Remaking of the Central City*. Oxford: Oxford University Press.

Lötsch, A. 1938. The Nature of Economic Regions. *Southern Economic Journal* 5: 71–8.

Luhmann, N. 1975/1991. Die Weltgesellschaft, in *Soziologische Aufklärung 2. Aufsätze zur Theorie der Gesellschaft*, edited by N. Luhmann. Opladen: Westdeutscher Verlag, 51–71.

Majoor, S. and Salet, W. 2008. The Enlargement of Local Power in Trans-Scalar Strategies of Planning: Recent Tendencies in Two European Cases. *Geojournal* 72(1–2): 91–103.

McNeill, D. 2005. In Search of the Global Architect: The Case of Norman Foster (and Partners). *International Journal of Urban and Regional Research* 29: 501–15.

Melbin, M. 1978. Night as Frontier. *American Sociological Review* 43(1): 3–22.

Meyer, J. 1951. The Stranger and the City. *American Journal of Sociology* 56: 476–83.

Meyer, J., Boli, J., Thomas, G.M. and Ramirez, F.O. 1997. World Society and the Nation-State. *American Journal of Sociology* 103(1): 144–81.

Müller, M. and Dröge, F. 2005. *Die ausgestellte Stadt. Zur Differenz von Ort und Raum*. Basel: Birkhäuser Architektur.

Mumford, L. 1961. *The City in History: Its Origins, its Transformations, and its Prospects*. New York: Harcourt, Brace & World.

Mumford, L. 1996: What is a City? in *The City Reader*, edited by R.T. LeGates and F. Stout. London and New York: Routledge, 184–8.

Park, R.E. 1915. The City: Suggestions for the Investigation of Human Behavior in the City Environment. *American Journal of Sociology* 20(5): 577–612.

Popp, M. 2004. Developing Shopping Centres in Inner-City Locations: Opportunity or Danger for Town Centres? *Die Erde* 135: 107–24.

Potuoğlu-Cook, Ö. 2006. Beyond the Glitter: Belly Dance and Neoliberal Gentrification in Istanbul. *Cultural Anthropology* 21(4): 633–60.

Priebs, A. 1992. Strukturwandel und Revitalisierung innenstadtnaher Hafenflächen – Das Fallbeispiel Kopenhagen. *Erdkunde(2)* 46: 91–103.

Reulecke, J. 1985. *Geschichte der Urbanisierung in Deutschland*. Frankfurt am Main: Suhrkamp Verlag.

Rofe, M.W. 2003. 'I Want to be Global': Theorising the Gentrifying Class as an Emergent Élite Global Community. *Urban Studies* 40(12): 2511–26.

Sassen, S. 1991. *The Global City: New York, London, Tokyo*. Princeton: Princeton University Press.

Schechner, R. 1993. The Street is the Stage, in *The Future of Ritual. Writings on Culture and Performance*, edited by R. Schechner. London and New York: Routledge, 45–93.

Simmel, G. 1903. Die Großstädte und das Geistesleben, in *Die Großstadt. Vorträge und Aufsätze zur Städteausstellung*, edited by T. Petermann. Dresden: Gehe-Stiftung zu Dresden, 185–206.

Sloterdijk, P. 2000. *Die Verachtung der Massen. Versuch über Kulturkämpfe in der modernen Gesellschaft*. Frankfurt am Main: Suhrkamp Verlag.

Stichweh, R. 2000. Zur Theorie der Weltgesellschaft, in *Die Weltgesellschaft. Soziologische Analysen*, edited by R. Stichweh. Frankfurt am Main: Suhrkamp Verlag, 7–30.

Stichweh, R. 2006. Zentrum/Peripherie-Differenzierungen und die Soziologie der Stadt. Europäische und globale Entwicklungen, in *Die europäische Stadt im 20. Jahrhundert. Wahrnehmung – Entwicklung – Erosion*, edited by F. Lenger and K. Tenfelde. Köln: Böhlau, 493–509.

Tallon, A.R. and Bromley, R.D.F. 2004. Exploring the Attractions of City Centre Living: Evidence and Policy Implications in British Cities. *Geoforum* 35(6): 771–87.

Thrift, N. 2005. But Malice Aforethought: Cities and the Natural History of Hatred. *Transactions of the Institute of British Geographers New Series* 30: 133–50.

Urry, J. 1990. *The Tourist Gaze, Leisure and Travel in Contemporary Societies*. London: Sage.

Williams, P. and Smith, N. 2007. *Gentrification of the City*. London and New York: Routledge.

Wirth, L. 1938. Urbanism as a Way of Life. *American Journal of Sociology* 44(1): 1–24.

Chapter 2

Planning Urbanity –
A Contradiction in Terms?

Loretta Lees

Planned communities designed with the goal of diversity … seem inevitably to attract accusations of inauthenticity, of being a simulacrum, rather than the real thing. Thus planners appear caught in an insoluble dilemma – either leave the market to take its course or impose an oxymoronic diverse order.

Fainstein (2005: 6)

[U]rban planning has finally made a critical shift towards urbanity. Yet, these outstanding steps in planning practice have been led by … ambivalent and often deceitful interpretations of urbanity, thus further reinforcing gentrification, spatial segregation and social inequalities.

Busa (2007: 2)

In what follows I expand on an editorial that I wrote recently (Lees 2010) that discussed, very briefly, the idea of planning urbanity. Over the past two decades or so 'urbanity' has been a constant presence in progressive urban planning debates. The divorce with modernist planning (a move away from the mono-functional urban district and the erosion of social interaction in suburban development) and the shift to an anti-modern urbanism seems now to be widely accepted and indeed advantageous to forward-thinking politicians, urban planners and developers. We appear to be in a new, postmodern era wherein the notion of urbanity is widely celebrated, advertised and squeezed into an often narrow, iconic vision by entrepreneurial city councils, planners and developers. The result is often ambivalence and unrealised potential (Lees 2003a) and the inauthentic urbanity of the gentrified inner city (see Zukin 2010). In 2008 I visited HafenCity in Hamburg, Europe's largest redevelopment project, where I was asked to address two basic questions: what is Urbanity? And – can/should we plan urbanity? I was forced to question, challenge and re-discuss urbanity, thinking through both the critical aspects and some of the possibilities that come out of the tensions, and at times contradictions, between urbanity and planning (Lees 2008a). For urbanity is about *unplanned* events and coincidences, it is about paradoxes and possibilities (Lees 2004), and as such it appears that planning urbanity is a contradiction in terms. Since that visit I have become increasingly interested in the concept of urbanity and indeed in the operationalisation of, and long-term outcomes of, the detailed plans for urbanity in HafenCity and in other regeneration projects in Europe and elsewhere. Wüst (2007) argues that urbanity is a myth and that it is likely to be instrumentalised by symbolic politics, that is using symbols, ideal visions and rhetoric in place of a concrete commitment to achieve a truly real (not fake or sanitised) urbanity. Indeed, urbanity is difficult to disagree with and its discursive flexibility has helped enrol otherwise hostile interests to the urbane projects envisioned (see Lees 2003a, 2003b).

What is Urbanity?

Urbanity is a somewhat elusive word. Most dictionaries don't even try to define it, instead they refer to an urbane quality or urban life. In the academic literature the term has a variety of different subdisciplinary framings – as Busa (2007: 3) points out:

> Among architects it may mostly reflect a mix of building types, of commercial offerings, and of architectural styles, whereas among urban planners and urban designers, urbanity may result from a mix of efficient public transportation, pedestrian-friendly streets and good social policies; for sociologists, anthropologists and geographers, tolerance and acceptance of the 'other', as well as mixed uses, and class or racial-ethnic heterogeneity, may represent the 'core' meaning of urbanity.

On the surface then it seems bizarre that the planning profession, developers and policy-makers talk about urbanity as if it were something definable and indeed attainable.

I would say that urbanity is defined precisely by the very elusive qualities that it contains. Urbanity in my mind is an aesthetic experience, involving but moving beyond the urban built environment. It is to some extent intangible. Quite simply urbanity is the unexpected that is produced by, or comes out of, the urban. Urbanity concerns emotions, randomness and chance, complexity and difference, contradictions, and the social. It is not readily tied down (cf. Amin and Thrift 2002).

The operational definition of urbanity that I'd like to begin with is based on the one put forward by Henri Lefebvre (see Merrifield 2006: Chapter 4). And I emphasise 'begin', because I feel that this can only be an ongoing process, in the same way that Henri Lefebvre argues that 'the urban' as a concept changes through time, so too I believe does 'urbanity' as a concept. Henri Lefebvre, like critics of modern architecture and planning, was critical of modern urbanism because it undermined urban life. Its rationality, comprehensiveness, scientific objectivity was/is at odds with urbanity. For Lefebvre urbanity was about encounter – the meeting of difference, of strangers in the city, it was about everyday life and play, the sensuality of the city. The city's built environment and its inhabitants were seen to be unique works in this process (the city as 'oeuvre'). This is a more fluid sense of urbanity than that caught by urban planning and urban policy – for here I see urbanity as something that is becoming, rather than being (cf. Amin and Thrift 2002: 80). As Pløger (2006: 393) states – urban space (like urbanity) cannot be seen as static, rather it must always be understood as being 'in process, always connected on one hand to flows of simultaneity and assemblies (lines of structural dependency) and on the other hand to unpredictable events, situations or mobility'.

Urbanity is both a socio-cultural construct and a built environment construct. As the former urbanity is a lifestyle or way of being linked to living or being in the city. This often gets translated (especially by politicians) as a form of civic culture that requires tolerance to difference in the city. Through this lens urbanity is a liberal middle-class rhetoric about quality of life in the city. As the latter urbanity is something that the built environment delivers through its very physical fabric – its architecture and public spaces, its densities and connections.

Schneider (1990: 2) cited in Wüst (2007) addresses both the socio-cultural and the built environment dimensions of urbanity, he says that on the one hand the idea of urbanity describes *an urban way of life* characterised by:

- a relatively differentiated structure of needs;
- the possibility to choose among different activities and experiences in different locations in urban space;
- the possibility to outdistance oneself and the ability to select specific social contacts;
- tolerance and mutual consideration;
- gender equality;
- a certain facility in dealing with different social roles and situations.

And on the other hand it encompasses features and criteria related to *the built environment of cities*:

- a wide offering of multi-faceted possibilities for cultural enrichment and educational opportunities;
- a wide offering of different living forms;
- harmonic proportions of the urban fabric;
- multi-functionality, mixed uses;
- clarity, order and functionality;
- pedestrian friendliness, proximity, public transit options;
- landmarks and places of local distinctiveness;
- a stimulating atmosphere of lively city culture, open-mindedness and multi-faceted social contacts.

We can see here that urbanity is more than the sum of its parts and few planners today would dispute the fact that urbanity is social and physical, but perhaps fewer would see it as something unfixed, as something in the process of becoming (cf. Pløger 2010).

A Planned Approach to Urbanity

Lefebvre argued that planning undermined urbanity. Yet planners today are striving to plan/create urbanity. How are they planning urbanity and what are the results of this?

With its emphases on functional and economic diversification and social diversity, postmodern planning has sought to juxtapose different residential, commercial and leisure users and uses of urban space in new hetero-spaces. Modernist mono-zoned spaces are being replaced by postmodernist hetero-zoned spaces, such as work–play, live–work and play–live, which offer a fine-grained dense network of uses. These newly designed hetero-spaces are celebrated as being more vibrant, liveable and uniquely urban. This replacement of spaces reflects the transition from a culture of production to a culture of consumption and reproduction, but deep tensions have arisen in these new hetero-spaces. These tensions have, however, given us greater scope for debate about what differences communities should tolerate in these new hetero-spaces (see Lees 2003a).

But interestingly the word urbanity seldom appears in planning documentation. When it does, it is used loosely in association with concepts like high-density, diversity and mixed use, the assumption being that if those elements are in place then urbanity will somehow follow. The assumption is that increased density will somehow produce/create urbanity, that community will 'be created "naturally" through the sustainable compact urban form that will densify people together cheek by jowl' (Lees 2003b: 78). Diversity and mixed use are part and parcel of the same thing, for example the British Urban White Paper (DETR 2000) uses the terms diversity and mixing interchangeably. The function and value of diversity and mixing, however, are simply taken at face value and are not questioned, they are seen to be 'the medicine for the centripetal forces of commercialised homogeneity, the centrifugal forces of suburban sprawl, and the exclusionary forces of housing segregation, both social and spatial' (Lees 2003b: 77).

Diversity is seen to produce difference, and living with difference is argued to be a positive experience that makes our lives richer and more meaningful, enabling personal development and even fulfilment, and perhaps most importantly (in multicultural liberal democracies) producing empathy for our fellow urbanites (cf. Sennett 1970, 1994).

Behind these assertions about cultural vitality (or *urban vitalis* as Pløger (2006) calls it) lie equally strong assertions about diversity and mixing being sources of economic renewal. There are two ideas here.

First, that cities innovate when people mix and mingle, sharing and combining ideas from different vantage points and traditions. For example, Richard Florida (2003, 2005) (who has written extensively on the so-called creative class and creative city) has produced a 'diversity index' linked to a 'gay index', his argument being that a city's openness to different kinds of people demonstrates its openness to new ideas and the new economy. Richard Florida's argument here is not new, it was made by urbanist Jane Jacobs in her later work, when she argued that diversity not only made cities more appealing but also acted as a source of economic productivity (see Fainstein 2005).

Second, densifying diverse categories of urban dwellers cheek by jowl is seen to promote social mixing, which is seen to create urbane citizens – who learn from each other and thus attain economic independence, creating Richard Florida's vision of a 'creative trickle down' (cf. Peck 2005: 766). Over the past decade the idea in British urban policy has been that if wealthier and more educated people mix with low-income groups their social and cultural capital will somehow trickle-down and help low-income groups into the job market and to gain economic independence. New Labour actively promoted mixed communities through urban renaissance initiatives in central urban areas (see Imrie and Raco 2003), in essence this was/is a form of state-led gentrification (Lees 2008b). We have yet to see what the current British government will do, but the signs are that it too wants to make us all middle class.

But as Holcomb and Beauregard (1981: 3) noted some time ago now 'Although it is often assumed that the benefits of revitalization will "trickle down" to the lower and working classes in a manner similar to that hypothesized for the housing market ... in fact they are almost completely captured by the middle and upper classes'.

Diversity and mixing then are seen to be a source of economic renewal and cultural vitality, but achieving the full potential of urban diversities requires more than just a celebration of abstract diversity, it requires a commitment to the broader concepts of tolerance and justice (Lees 2003b: 78). This tallies with Leonie Sandercock (2004: 134, italics added) who defines planning as an '*always unfinished* social project whose task is managing our coexistence in the shared spaces of cities and neighbourhoods in such a way as to enrich human life and to work for social, cultural, and environmental justice'. Here, like urbanity, planning itself is in the process of becoming.

Some Examples of the Contradictory Outcomes of Planned Urbanity

The Case of Portland, Maine, USA: Planning Urbanity Leads to Ambivalence

Urban renaissance initiatives have been deeply ambivalent about urban diversity – they promote diversity at the same time as promoting social controls that limit diversity, as the large number of academic accounts on displacement and domestication in urban public space reveal. One such account is the research I undertook on planning for diversity in Portland, Maine, a small city on the north-east coast of the United States (see Lees 2003a). Over the past 35 years Portland has been transformed from a de-industrialised port city into a major tourist attraction and hub for commercial activity. A central plank of Portland's plans to make itself a more prosperous and liveable city has been the promotion of diversity. Successive planning documents have tried to break with the functional city of modernism and replace it with a new diverse, pastiche city.

Portland's 1988 downtown plan – *A Celebration of Urban Living* – imagined diversity as synonymous with urban vitality and called for a multi-functional diversification of downtown through the simultaneous economic, functional, cultural and spatial diversification of downtown and downtown living. 'While this wide range of uses threatens to make the term incoherent, it is also key to the appeal and power of diversity in planning discourse. Like motherhood and apple pie, diversity is difficult to disagree with. Janus-like, it promises different things to different people' (Lees 2003b: 622). Wüst (2007) argues similarly that the word urbanity was 'predestined for the political rhetoric of urban development'. In its various guises diversity offered a harmonious, win-win picture of Portland's future downtown development that could command the support of a heterogeneous coalition of small business owners, corporate interests, arts and educational institutions, municipal officials and Portland residents. Indeed, the subsequent 1991 *Downtown Vision Plan* in which diversity assumes an almost iconic status made strategic use of the idea of diversity to build support for its vision of both a '*Downtown for People*' – 'where people of all ages and all socio-economic groups find an exciting, friendly and compassionate atmosphere ... on the sidewalks and open spaces throughout the downtown; and where people want and are encouraged to come together' – and a '*Downtown for Opportunity*' – 'where a bustling office and retail economy combines with a thriving and diverse cultural, entertainment, and visitor economy to provide a prosperity shared by the whole community'.

But, from day one there were tensions between the 'downtown for people' and the 'downtown for opportunity'. In the new hetero-spaces of downtown Portland, profitable bars and nightclubs attracted college kids and young urban hipsters whose night-time noise offended nearby residential occupants. Likewise the daytime behaviour of youths – skateboarding and hanging out in front of stores – disturbed residents and shopkeepers. Indeed, one of the most prominent manifestations of the tensions over diversity in downtown Portland was a long-running conflict between youth and local businesses. Quite simply the youth were seen to be undesirable bodies on the streets and in the public spaces of the new cultural and commercial spaces of downtown Portland. The local businesses said that they were putting off customers. The response from the City was attempts to push the kids out of the newly diversified spaces of downtown by installing speakers playing light classical music, police swoops, and opening a police station and introducing a farmers' market two days a week in Monument Square where they hung out, creating City by-laws for two of the small downtown plazas or parks where the kids hung out disallowing access after 10 p.m., and finally they built a skate park on the fringes of the city hoping the kids would move there. Where diversity was seen to be a threat to the local economy eviction and marginalisation from the public spaces of the city ensued.

Ironically, efforts to foster urbanity often subvert that very goal, as efforts to secure urban space for public culture stifle diversity and vitality (Lees 1998: 251). Such tensions and conflicts were suppressed by the functional and spatial segregation of the modern city in which alternative activities and social groups were kept apart in the interests of social order and economic efficiency. Yet these new plans for urbanity are in danger of seeking to resuscitate the defunct or merely hibernating ambitions of modernity by cleansing these multi-functional spaces of the other faces of diversity through heavy-handed policing. The lesson here is that we need to go beyond optimistic hopes for a functional mix and be clear as to what we mean by diversity and how we will deal with the fallout from diversity. Tolerance to difference is one of the main features of urbanity, and it is also the core value in Sandercock's (2003, 2004) definition of planning.

The Case of the Aylesbury Estate, London, UK: Planning Urbanity Leads to the Resuscitation of the Ambitions of Modernism and to State-led Gentrification

It is interesting that some of the larger redevelopment projects in Europe, premised on ideas about diversity and mixed use, seem to be reinventing the modernist comprehensive redevelopment schemes of the past. Take the Aylesbury Estate in central London (see Davidson and Lees 2010), which is the largest public housing estate in Europe and is in the process of being demolished and replaced with mixed income new-build housing (although recently announced government cutbacks are now affecting the scheme[1]). The Aylesbury, constructed over a 10-year period from 1967, was at the time one of the most ambitious post-war developments by any London borough. The estate is a stereotypical example of modern architecture – it is a mesh of panel system built towerblocks and lowrise flats with concrete walkways. It has a reputation for poverty and crime. The modernist vision for the Aylesbury was of an egalitarian society created through architecture and planning, the postmodernist vision is for an inclusive, mixed community. These visions are both similar but quite different.

The demolition and displacement that occurred with the 1960s slum clearances for the initial construction of the Aylesbury is no different to the large-scale demolition and displacement that is taking place now. The arguments put forward then and now are similar – poor quality housing, poverty, crime, etc.; but the means to attain redevelopment has changed significantly – from centrally planned mass production to a public/private partnership that plays off policy rhetoric in the UK on mixed communities. On the day after New Labour's general election victory in 1997 Tony Blair made a surprise visit to the Aylesbury where he made a speech highlighting the estate's residents as Britain's 'poorest' and the 'forgotten'; many of whom 'played no formal role in the economy and were

---

1   With the first residents preparing to leave their flats for new, sustainably-built homes the rest of the redevelopment is suddenly in jeopardy after more than £180m of central government funding was withdrawn without notice.

dependent on benefits'. Very quickly afterwards the Aylesbury was given New Deal for Communities (NDC) status and studies began on how the estate could be redeveloped. The plan is to break down this large and socially homogenous monolith of public housing and to replace it with a more mixed and diverse or heterogeneous community. As Damaris Rose (2004: 281) argues: 'Since the image of the "liveable city" has become a key aspect of a city's ability to compete in a globalized, knowledge-based economy (Florida 2003), post-industrial cities have a growing interest in marketing themselves as being built on a foundation of "inclusive" neighbourhoods capable of harmoniously supporting a blend of incomes, cultures, age groups and lifestyles'.

The current strategy for the demolition and rebuilding of the Aylesbury Estate (The Aylesbury Estate: Revised Strategy) lists the construction of 3,200 private new-build homes and 2,000 social rented new-build homes. This fulfils the UDP requirement for 40% social housing. They are seeking to demolish the vast majority of the Aylesbury (despite much of it being structurally sound) and to create a new-build, mixed community. This plan does not acknowledge the current mix already in the area (which is already very socially and ethnically diverse), nor does it address issues of social sustainability. It will also undermine the estate functioning as a reception area for people seeking asylum and will displace this elsewhere. 'If the tenants are not being socially cleansed, they are being treated like guinea pigs' (Matt Weaver, *Guardian*, 22 September 2005). The new community, a kind of sim city, will be twice the current density with the majority of the inhabitants being the sought-after middle classes. The plans for the Aylesbury are nothing less than a form of state-led new-build gentrification that has refused to consider the rising inequality that emerges when urban policies sold to us through the neutralising rhetoric of social mixing/diversity aim to attract and retain the middle classes.

Planned Urbanity in HafenCity

HafenCity is a complex urban redevelopment project, which when completed will nearly double the size of downtown Hamburg. Significantly, its development team see it as a point of departure from other redevelopments because they are seeking to create a new kind of urbanity that will be the model for the European city of the 21st century that 'conceives and realises new forms of inner city coexistence'. 'Urbanity' is a technical term in the glossary for HafenCity and is defined thus: 'The metropolitan feel of an urban space. Urbanity basically arises from an appropriate urban structure made up of a variety of land uses in and outside buildings, as well as the presence of a large number of people of different types' (www.hafencity.com/en/glossary-s-z.html~urbanity). The development team has made great strides to create what they call 'a new kind of urban experience' where the most varied uses and users come together to create 'an intensive form of inner city community'. Twelve thousand residents will share HafenCity with 40,000 workers and 80,000 visitors. Myriad uses are concentrated in a limited space, homes with offices, shops, cafes, etc. The target market is equally varied: singles, young families, empty nesters, retirees, gay couples, etc. HafenCity seeks to bring together 'people, milieus, lifestyles and interests that might otherwise never come across each other'. The fine-grained mix is not just at the neighbourhood level, but also at the level of individual buildings. The ground floor stories of buildings are publically accessible and there are numerous opportunities for social encounters: 'The public urban spaces with their different characters invite the most varied uses, while public right of access on private ground between buildings is also widespread ... private spaces in Überseequartier even allow journalistic reporting, political advertising, demonstrations and begging'. As Bruns-Berentelg (forthcoming) makes clear, the 'urbanity' planned for HafenCity is not a distributive form of social mix for residential households (because this did not exist in German policy – so there was no funding), indeed he is not convinced that the addition of low-income housing would contribute directly to an emancipatory model of urbanity in the new city. Rather the planned urbanity in HafenCity has meant pre-selecting some residents for their predisposition to the communitarian notions of urbanity that HafenCity advertises and designing the development so that the physical and institutional environment creates and offers possibilities for urban social encounter. This is a performative urbanity, an 'urbanity-in-becoming' (cf. Pløger 2010): 'Social mix amongst residents and the capacity for encounter in public-social places are mutually dependent: public places of encounter must be socially available for collective urbanistic purposes in order to have emancipatory effects, and the urban role of residents – rooted in the city through the private character of residence – can unfold meaningfully and to a considerable extent in such public social places' (Bruns-Berentelg forthcoming).

HafenCity Hamburg GmbH are so committed to creating urbanity that they are funding ethnographic studies of whether the uses of the public spaces they have designed are living up to their expectations, they are also undertaking longitudinal studies of social mixing in HafenCity from the time residents first move in. These research endeavours are aided by the new HafenCity University on the site, a unique university that specialises in teaching and research on the city and the future of cities. They have already begun to answer Pløger's (2010: 336) call for urban planners who want to plan urbanity to do detailed analysis of micro-social forces – to be the 'wandering planner' who 'botanizes'. Pløger adds that this work 'requires a sensitive dialogue with subjects of "otherness" and the feelings of being different or an-other, which might require planning institutions to work with "not-understandable" ways of living and doing ... This would be a real bottom up planning from where planners may be able proactively to anticipate changes in urban life before they are manifest'. There is scope for this yet at HafenCity.

It is ironic then that given HafenCity is so committed to 'urbanity', to open public space, to mixed uses and social mixing, all the things that critical urban researchers have been advocating for some time, that HafenCity has been tarred with the same brush of negative media coverage (e.g. Oehmke 2010) and academic coverage (e.g. NION Brand Hamburg 2010) that the Florida-inspired (2003, 2005) 'Hamburg City of Talent' (the Brand Hamburg) has. Yet if one looks at the ideas behind HafenCity they are not Floridean, they are not merely an obsession with attracting the creative class (see Peck 2005), they are the ideas of critical planners and urbanists. Critical urbanists need to be careful and do their homework – where anti-gentrification activists in Hamburg, led by the Right to the City group, discuss Henri Lefebvre's 'right to the city', HafenCity tries to encapsulate Henri Lefebvre's notion of 'urbanity'. Jürgen Bruns-Berentelg, the CEO of HafenCity Hamburg GmbH is a trained geographer and has a substantive knowledge of contemporary critical urban theory, which he has used in his detailed social plans for HafenCity. Indeed HafenCity counters the charge of gentrification thus:

> Occasionally the term gentrification is used in connection with HafenCity because the housing being built is of high quality and negative repercussions are feared. But this criticism is not valid. The phrase was coined in 1964 by Ruth Glass to describe the crowding out of households from their traditional living quarters by new groups of higher-income residents who can afford to invest in the buildings, thus increasing property values in the neighborhood. This process, to be observed in the Hamburg districts of St. Pauli, St. Georg or Schanzenviertel, is not taking place in HafenCity; as the district never boasted any housing, no established residents exist to be squeezed out of their homes. Indirect gentrification in new-build areas is not evident either. HafenCity is separated by the Speicherstadt from other inner-city residential quarters and, because they are subsidized, homes close to HafenCity are better protected against increasing rents than old buildings in other inner-city locations. Direct gentrification is not to be expected because the volume of new-builds outside HafenCity within the surrounding inner city is small. On the other hand, it is important that a large quantity of high-quality house building is developed in HafenCity. It increases the proportion of high-value homes which is below average in the inner city and in this way upgrades the quality of social and recreational infrastructure for city-center residents surrounding HafenCity. At the same time, the development of expensive homes in HafenCity lessens the pressure for redevelopment in areas close to the inner city. So HafenCity is actually making a contribution towards less gentrification in Hamburg. (www.hafencity.com/en/faq-concepts-planning/is-hafencity-causing-gentrification-.html)

The case of HafenCity is a complex one, unlike other urban renaissance initiatives it is not ambivalent towards diversity, indeed it has been planned to make sure that diverse uses and people co-exist, but as yet the verdict is out on the frictions and conflicts that will inevitably exist between different uses and different types of people. HafenCity could be charged with reinventing the modernist comprehensive redevelopment (renewal) schemes of the past with all their echoes of social engineering, a charge that other large redevelopment projects premised on ideas about diversity and mixed use face, for urbanity in HafenCity is being socially engineered. And like in other cases (see Davidson and Lees 2010), the *mixed* district of HafenCity (whose residents are all middle to upper class) could be seen as a form of state-led new-build gentrification, but given my knowledge of the critical thought that has gone into this redevelopment the verdict is still open on this. I hope my writing here will persuade HafenCity to take more careful account of their planned urbanity, so that it properly includes the poor in society – those economically marginalised populations who are steadily losing their place (and homes) in European inner cities and elsewhere (Lees et al. 2008). And this means more than merely adding in some social housing as HafenCity has now been required to do (see Bruns-Berentelg forthcoming). Moreover, as Busa (2007: 8) states: 'Developing an appropriate physical setting for a heterogeneous urbanity ... can go only so far in the generation of a just city. Most crucial is a political consciousness that supports progressive moves at national and local levels towards respectfulness of others and greater equality'. Local politics in Hamburg and national politics in Germany will determine the measure of that respectfulness, equality and inclusiveness.

Conclusion

The utopian impulse at the heart of so many experiments in city-building has always proved disappointing, if not downright disastrous, in the actual flesh and stone. Much has been written about why this is so – perhaps enough to discourage any further attempts at utopian thinking about the city. But the utopian impulse is, and will hopefully remain, an irrepressible part of the human spirit (Sandercock 2003: 2).

As I have shown above the postmodern urbanity that is planned for downtowns and urban communities nowadays often leads to displacement, social exclusion and social segregation (all typical features of gentrification). Yet, as I have also argued, the rhetoric of urbanity – whether it be for a diverse downtown, a mixed urban community, or a space of social encounter is very persuasive. I have come to the conclusion that if socially just goals of urbanity are to be possible in practice we need to match up the rhetoric of urbanity more readily with the reality (the outcome), in order to allow planned urbanity to produce a city for the many rather than the few (Amin et al. 2000). To do this we need an elastic but targeted definition of urbanity, elastic enough to allow the unexpected which may emerge (the becoming) to be drawn under its umbrella, but at the same time targeted enough to give urban planners and developers something to work with. Urbanity is an ambivalent and contradictory process, but by confronting it as such, perhaps we can take better advantage of its positive side as a social, aesthetic and moral form, whilst mediating its more negative side as a producer of social inequality and problematic social controls. As David Harvey (1992: 599) argues: 'just planning and policy practices must confront the phenomenon of marginalization in a non-paternalistic mode and find ways to organize and militate within the politics of marginalization in such a way as to liberate captive groups from this distinctive form of oppression'.

Is it possible to construct a concept of 'urbanity' on which it is possible to build a theory and practice of urban design and planning that will be useful not just now but in the future? I'm not sure, but I feel planners must try, as I begin to do in the table below.

Table 2.1   An elastic but targeted definition of urbanity for planners

High-density/densification

Multi-functionality, mixed uses

Possibilities for cultural enrichment and educational opportunities

Possibilities for different forms of living

Possibilities for different experiences in urban space

Reinstating the street as a pedestrian friendly space

Public transit options

Emphasis on landmarks and places of local distinctiveness

Emphasis on a lively city culture

Emphasis on tolerance, mutual consideration and open-mindedness in urban public spaces

Allow 'visible' spaces for the poor, socially marginal and/or deviant

Relatively equal numbers of different socio-economic groups

Strategies in place to prevent urban sanitisation/gentrification

Source: Drawing on Wüst (2007).

We cannot plan the unexpected, all we can do is to try and plan those elements of urbanity that are less elusive, and to make sure that we try hard not to sanitise the urban in the process. We need to allow space for the threatening, menacing city of disorder that is painful, shocking but also so alluring. The gritty urbanity that ironically attracted pioneer gentrifiers leading down the line to a sanitising form of gentrification. It seems to me that this is the urbanity that we are currently destroying to our cost with a form of sim-urbanity – a sterile form of the urban where nothing is unexpected, where adventure, possibility, urban stimuli and sensory experience are both stifled and scripted, not naturally occurring. Andy Merrifield says that we need to capture the ambiguity of urbanity, keep the city's scars in view to enable a politics of hope. Drawing on Henri Lefebvre, he argues that we need to put in place 'the right to the city' for all – 'the right to participate in urbanity, the right to appropriate the city not merely as an economic unit, but as a home and as an expression of lived experience' (2002: 156). In an increasingly urban world such utopian hopes for the city are more urgent than ever before (Lees 2004).

References

Amin, A. and Thrift, N. 2002. *Cities: Reimagining the Urban*. Cambridge: Polity Press.

Amin, A., Massey, D. and Thrift, N. 2000. *Cities for the Many not the Few*. Bristol: The Policy Press.

Bruns-Berentelg, J. (2011). Social Mix and Encounter Capacity – A Pragmatic Social Model for a New Downtown: The Example of HafenCity Hamburg (Germany), in *Mixed Communities: Gentrification by Stealth?*, edited by G. Bridge, T. Butler and L. Lees. Bristol: Policy Press, 69–94.

Busa, A. 2007. Celebrations of Urbanity – Editorial, *The Urban Reinventors* [Online] 2. Available at: www.urbanreinventors.net/2/busa/busa-urbanreinventors. pdf [accessed: 8 February 2011].

Davidson, M. and Lees, L. 2010. New-Build Gentrification: Its Histories, Trajectories, and Critical Geographies. *Population, Space and Place* 16(5): 395–411.

DETR 2000. *Our Towns and Cities – The Future. Delivering an Urban Renaissance.* [Online: Department for Communities and Local Government, London]. Available at: www.communities.gov.uk/documents/regeneration/ pdf/154869.pdf [accessed: 8 February 2011].

Fainstein, S. 2005. Cities and Diversity: Should we Want it? Can we Plan for it? *Urban Affairs Review* 41(1): 3–19.

Florida, R. 2003. *The Rise of the Creative Class: And How it's Transforming Work, Leisure, Community and Everyday Life*. New York: Basic Books.

Florida, R. 2005. *The Flight of the Creative Class: The New Global Competition for Talent*. New York: Harper Business.

Harvey, D. 1992. Social Justice, Postmodernism and the City. *International Journal of Urban and Regional Research* 16(4): 588–601.

Holcomb, H.B. and Beauregard, R.A. 1981. *Revitalizing Cities*. Washington, DC: Association of American Geographers.

Imrie, R. and Raco, M. 2003. *Urban Renaissance? New Labour, Community and Urban Policy*. Bristol: Policy Press.

Lees, L. 1998. Urban Renaissance and the Street. Spaces of Control and Contestation, in: *Images of the Street. Planning, Identity and Control in Public Space*, edited by N.R. Fyfe. London, New York: Routledge, 236–53.

Lees, L. 2003a. The Ambivalence of Diversity and the Politics of Urban Renaissance: The Case of Youth in Downtown Portland, Maine. *International Journal of Urban and Regional Research* 27(3): 613–34.

Lees, L. 2003b. Visions of 'Urban Renaissance': The Urban Task Force Report and the Urban White Paper, in *Urban Renaissance? New Labour, Community and Urban Policy*, edited by R. Imrie and M. Raco. Bristol: Policy Press, 61–82.

Lees, L. 2004. *The Emancipatory City? Paradoxes and Possibilities*. London, Thousand Oaks and New Delhi: Sage Publications.

Lees, L. 2008a. Planning Urbanity, Keynote speech presented to the *Symposium – Planning Urbanity: Life/Work/Space in the New Downtown*, HafenCity, Hamburg, Germany, March.

Lees, L. 2008b. Gentrification and Social Mixing: Towards an Urban Renaissance? *Urban Studies* 45(12): 2449–70.

Lees, L. 2010. Planning Urbanity? *Environment and Planning A* 42(10): 2302–8.

Lees, L., Slater, T. and Wyly, E. 2008. *Gentrification.* New York: Routledge.

Merrifield, A. 2002. *Dialectical Urbanism: Social Struggles in the Capitalist City*. New York: Monthly Review Press.

Merrifield, A. 2006. *Henri Lefebvre: A Critical Introduction.* New York: Routledge.

NION Brand Hamburg 2010. Not In Our Name! Jamming the Gentrification Machine: A Manifesto. *City* 14(3): 323–5.

Oehmke, P. 2010. Squatters take on the Creative Class: Who has the Right to Shape the City? *Spiegel Online International* [Online, 7 January]. Available at: www.spiegel.de/international/germany/0,1518,670600,00.html [accessed: 14 January 2010].

Peck, J. 2005. Struggling with the Creative Class. *International Journal of Urban and Regional Research* 29(4): 740–70.

Pløger, J. 2006. In Search of Urban Vitalis. *Space and Culture* 9(4): 382–99.

Pløger, J. 2010. Urbanity, (Neo)Vitalism and Becoming, in *The Ashgate Research Companion to Planning Theory*, edited by J. Hillier and P. Healy. Aldershot: Ashgate, 319–41.

Rose, D. 2004. Discourses and Experiences of Social Mix in Gentrifying Neighbourhoods: A Montreal Case Study. *Canadian Journal of Urban Research* 13(2): 278–316.

Sandercock, L. 2003. *Cosmopolis II: Mongrel Cities for the C21st*. London: Continuum.

Sandercock, L. 2004. Towards a Planning Imagination for the C21st. *Journal of the American Planning Association* 70(2): 131–41.

Schneider, H. 1990. Urbanität als Planungsfaktor. *Schweizer Ingenieur und Architekt* 1(2): 22–3.

Sennett, R. 1970. *The Uses of Disorder*. New York: Knopf.

Sennett, R. 1994. *Flesh and Stone: The Body and the City in Western Civilization.* New York Norton.

Wüst, T. 2007. The Urbanity Myth [Online] 2. Available at: www.urbanreinventors. net/2/wuest/wuest-urbanreinventors.pdf [accessed: 8 February 2011].

Zukin, S. 2010. *Naked City: The Death and Life of Authentic Urban Places.* Oxford: Oxford University Press.

Chapter 3

Public Spaces for the 21st Century

Jan Gehl

A Public Life Department?

All cities have traffic departments that carefully collect data concerning how the traffic is developing. Every year traffic counts are made and prognoses worked out. Based on all this good work it is easy to discuss traffic issues and future traffic policies for the city. All this makes traffic very *visible* in the planning process. Traffic is a serious matter!

In sharp contrast to this is the handling of people in the cities (Gehl and Gemzøe 2001). It appears that hardly any cities have a department for pedestrians and public life, and hardly any cities have data concerning how people are using the city. Are there more pedestrians than 10 years ago – or fewer? And is this good or bad? Do the streets, squares and the parks work well for people? Or not at all? In most cities there are many good intentions, but seldom facts and evidence to qualify the discussion about the users of the city. For these reasons the people using the city are generally overlooked in the planning process. The people using the cities are in most cases almost *invisible*!

From Necessity to Option: The Changing Character of Public Life

Automobile traffic has come into our cities in great numbers from the 1950s onwards, and this new phenomenon has been attended to by the traffic professionals. And, of course, this is fine in itself!

Pedestrians and public life, on the other hand, have been around for thousands of years. On the surface this appears as an unbroken tradition of activities, but in reality during the past 100 years a number of dramatic changes in the character of public life have taken place, all of which makes concern and careful planning for people in the city urgently needed today.

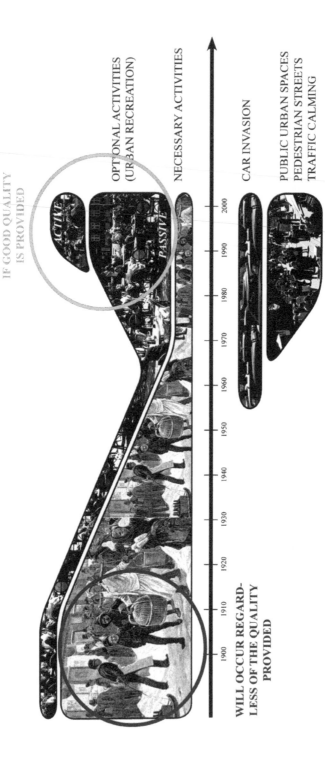

Figure 3.1   Illustration from *New City Life*
Source: Gehl and Gemzoe (2006).

Comparing street scenes from 100 years ago with present-day street scenes an obvious change in the volume and character of the public life stands out. In the bustling street scenes from around 1900 nearly everyone is engaged in some type of necessary activity. People are present because they have to be, regardless of the quality of the environment. Use of public spaces was an important part of daily life, and the spaces were filled to overflowing with all kinds of activities – and in the process the public spaces also functioned in the ever-present role as meeting places for people. This type of streets scene – filled with people who are, day after day, compelled to use the public space – is still to be found today in many less economically developed countries.

Street scenes from our present-day societies with developed economies show a distinctly different use pattern. Few people are present today because they are forced to be. Some walking to and from parking lots or to and from work or public transport would be in this category, but many others have alternative options for transport, for reaching services and for shopping. In short, very few people today have strong reasons to walk or to be present in the public realm.

In city areas depending heavily on public transport, such as Manhattan, central London or Tokyo, great crowds of pedestrians can still be found. They are still out there walking because they have to. But in many other cities – where no comfortable walking facility for people is provided – you will find hardly anyone in the streets. The cities become deserted and empty. This sad state of affairs can be found in quite a number of cities in the United States. And the phenomenon is spreading to other countries with similar ideologies and car-dependent lifestyles.

Fortunately by now another category of city can be found – the 'reconquered cities'. In these cities incentives to walk and to use the public realm have been carefully made by providing the quality and ambience, which today are essential to address the shift from necessary to optional activity patterns.

The overall picture of the lively present-day public spaces reveals that most of the people are present because they want to be. They have decided to come and to stay because the public spaces continue to offer valuable opportunities for them.

The optional character of most public life activities in present-day cities however places very high demands on the quality offered by the public realm. If the quality is missing, people will not come; they have many other interesting options open to them. However if the public spaces are well placed, well designed and inviting, evidence from all over the world points to the fact that people continue to appreciate public life in the public spaces. If the invitation is right people will come!

Thus if lively, attractive and safe cities with active streets, squares and parks are to be realized in the 21st century, then good provision for pedestrian activities is vital (Gehl 2006, 2007).

Winning Back Public Space – Winning Back Public Life

For the past 40 years the issues of public space quality and conditions for public life have been debated, and the theories tested in numerous projects and city districts. The evidence collected by now is quite overwhelming. From the largest of cities to the smallest of individual spaces the interrelationship between quality provided and the volume and character of people's activities has been firmly established. If public spaces of good quality are provided then people in present-day society will use them – because they want to. This can be illustrated by the success stories of cities such as Barcelona, Lyon, Strasbourg and Portland, USA. And many more humble cities and places confirm it.

The quality issues to be looked into are by now well established, and also methods on how to turn a city around have been developed and documented. The virtues of a systematic and careful approach have been demonstrated.

A method for improvements pioneered in Copenhagen and by now used widely – the latest examples being central London, Melbourne, Sydney and New York – is to survey systematically how people use the city. Based on this knowledge (diagnosis) a cure for improvements can be worked out. Books such as *Public Spaces – Public Life* (Gehl and Gemzøe 1996) and *Towards a Fine City for People* (GEHL-Architects 2004) describe in detail how such a method can be applied and which tools to use. Further, *Public Spaces – Public Life* describes the amazing increase in the use of public spaces, following the many quality improvements, which has occurred in the course of the past four decades.

Copenhagen has for many years served as the first city where systematic recordings of the public spaces and their use have been instrumental for discussions, policies and improvemennts. The public space–public life surveys have served to establish knowledge and have given the planners and politicians confidence to proceed with the city improvements, given that the previous interventions have been documented as being successful.

Subsequent surveys carried out in a number of other cities – from major cities such as Stockholm, Oslo, Riga, Edinburgh, Rotterdam, Wellington, Adelaide, Perth and London to humble provincial cities – have over the years further demonstrated the usefulness of public life surveys as a basis for improvement strategies, and impressive increases in the volumes and character of the public life have been documented.

For a number of years it has therefore been accepted in urban planning theory that improvements to the pedestrian environment might result in a more lively and attractive city, where more people would like to walk and spend time in the city. Evidence from various cosy old European cities with crooked streets and romantic buildings have been plentiful; Lyon, Barcelona, Strasbourg, Freiburg and Copenhagen, just to mention a few, tell this story.

Figure 3.2   Federation Square Melbourne – a contemporary city space for people

Amazing Melbourne

Melbourne now adds a new dimension to these tales. Melbourne is a young colonial city, with wide, straight streets and no 'built-in', inviting urban squares from the outset. Furthermore, it is a city studded with uncoordinated high-rise office developments dating from the 1960s and 1970s, leading to it being characterised in the local newspapers by 1978 as a monofunctional, empty and useless city centre. A 'doughnut', the newspapers called it by this time. A nondescript place with nothing in the centre. A city with no beating heart! Many cities across the New World – and many new city districts in the Old World – will fit this description. And in most of these places, the car continues to be king and the 'doughnut syndrome' still prevails.

This, however, is not the case in Melbourne anymore. A carefully planned and executed process for turning the city into a people-oriented city has been gradually implemented over the past two decades. Of all the things a city can do to improve the environment for people, Melbourne has done most everything. More residents, more students, more pedestrianised streets, new squares, lanes and parks, wider sidewalks, quality materials, active building frontages, fine furnishings and several art programmes. Also sustainability issues such as greening the city and upgrading the public transportation systems and the bicycle infrastructure have been systematically addressed.

Most of this has been accomplished over a short span of years and in the process life in the public realm of central Melbourne has changed dramatically. A public space–public life survey carried out in 2004, and conducted along the lines of a previous survey from 1993, shows remarkable changes concerning how people now use the city. Many more people are walking the streets. On weekdays some 40% more and in the evenings twice as many as in 1993. And many more people come to town to promenade or spend time enjoying the city, the sights and the number one city attraction: other people. Compared to the situation in 1993, some 3–4 times more people are found standing, sitting and otherwise using the squares, the parks, the street benches and many new café chairs.

A city scolded in late 1970s for having 'an empty useless city center' has in 25 years been turned around to become a vibrant, charming 24-hour city centre – more lively, more attractive and safer than most other cities found in the New World. An almost Parisian atmosphere yet 'Down Under!'.

The Melbourne story gives hope to cities in all parts of the world, and an added incentive for this is the well documented growth in the city economy arising from increasing popularity and attractiveness of the public realm.

Street scenes in Melbourne now have an almost Parisian atmosphere with wide sidewalks, ample furnishing and new street trees.

Figure 3.3   Melbourne street zones

References

Gehl, J. 2006. *Life between Buildings. Using Public Space.* Copenhagen: Danish Architectural Press.

Gehl, J. 2007. Public Spaces for a Changing Public Life. *Topos* 61: 16–22.

Gehl, J. and Gemzøe, L. 1996. *Public Spaces – Public Life.* Copenhagen: Danish Architectural Press.

Gehl, J. and Gemzøe, L. 2001. *New City Spaces.* Copenhagen: Danish Architectural Press.

Gehl, J. and Gemzøe, L. 2006. *New City Life.* Copenhagen: Danish Architectural Press.

GEHL-Architects 2004. *Towards a Fine City for People – Public Spaces and Public Life-London 2004* [Online]. Available at: www.gehlarchitects.dk/files/pdf/London_small.pdf [accessed: 8 February 2011].

Chapter 4

Waterfront Redevelopment: Global Processes and Local Contingencies in Vancouver's False Creek

David Ley

Waterfront redevelopment is now a widely replicated urban process that has transformed old port cities in Europe, North America, Asia and Australia, with such celebrated examples as the London Docklands, Hamburg's HafenCity, Baltimore Harbor, River City 21 in Tokyo, and Sydney's Darling Harbour. The need to establish new land use functions and economic returns on large but under-utilised central city property is equally felt in smaller cities and particularly imaginative and widespread harbour transformations may be found in Cardiff Bay in South Wales, Malmo in Sweden and Vancouver's False Creek. A fundamental shift lying behind these projects has been the transition from an industrial to a post-industrial society in many cities of the global north, giving rise to new downtown labour markets and inner-city housing markets (and lifestyles) that confound the traditional models of urban land use and social areas (Ley 1980).

In these older models an industrial area around downtown was located around rail yards and accessible water bodies, with adjacent rings of working-class housing serving these employment sites. But as Ilse Helbrecht and Peter Dirksmeier suggest in the introduction to this volume, a new urbanism is unfolding, of which the redeveloped waterfront is perhaps the most articulate example. The changes in urban structure and land use are both dramatic and rapid, covering large areas, requiring significant public–private investment, and intimating rapid landscape transition in the course of a decade or even less. In some instances like London's Docklands and Hamburg's HafenCity a secondary downtown is under construction, in other cities a specialised inner-city district is embellishing the residential, commercial and leisure options of the existing central city. The global repetition of these districts indicates at once that processes exceeding the regional and national scales are at work; indeed in some cases serial reproduction is taking place, as a successful model is imitated elsewhere. As the detailed example from Vancouver's False Creek will show later, both historical and geographical contexts at several scales are significant in the landscape manifestation of these processes in individual cities.

Transitions and Interpretations

If the Beatles had looked south from the Reeperbahn music clubs on their visits
to Hamburg in the 1960s they would have seen a busy working port much like
the Merseyside waterfront in their home town of Liverpool. In both Hamburg and
Liverpool in the 1960s there would have been thousands of dock-workers loading
and offloading ships, with other blue-collar navvies engaged in transport and
maritime services. Large warehouses and processing industries would round out
this concentrated economic activity. Today that busy world is largely gone, replaced
by downstream container terminals and port automation. By 2008 Liverpool was
converting itself into the European Capital of Culture, following the example of the
lapsed industrial port of Glasgow, with festivities launched in January that year by
none other than former Beatle Ringo Starr. The 19th-century industrial waterfronts
had grown silent while they awaited a new calling. In Cardiff Bay, where the coal
docks at the turn of the 20th century claimed to handle a greater annual cargo
tonnage than New York, a recent barrage across the Bay has converted the port into
a freshwater lake. Like waterfront developments elsewhere, the former docks have
been reconfigured, and now contain a mixed land use of apartments, offices, shops,
restaurants and cultural and leisure spaces.

How do we interpret the contemporary synthesis of land use and lifestyle that
comprises the landscape of the new inner city? In many respects it reflects the
new service economy, as jobs have been created in professional and managerial
fields while blue-collar positions have declined. The new inner city is the home
of Richard Florida's so-called creative classes. The loosened grip of the past
means that industry's polluting externalities are outlawed; the new inner city is
green in appearance and sometimes also in political sentiment. There is a new
emphasis on visuality and aesthetics as part of a liveable city: design is prioritised
as part of a scrupulous programme of socio-spatial engineering. The past itself is
aestheticised, selectively preserved as a cultural amenity, as dockside cranes are
re-imagined as outdoor sculptures, industrial buildings are retained as heritage
structures; museums and aquaria re-present the maritime world for contemplation.
Convivial and lively public spaces shape a festive core, created for arts and leisure
activities, though places of worship, a primary public space in the old inner
city, are commonly absent. Families are welcome daytime visitors, but at night
apartments are usually occupied by adults, in one- or two-person households.
A challenge facing the public–private partnerships responsible for these new spaces
is the participation of poorer families, and their claim to these districts. How broad
and deep is the public sphere about which Jürgen Habermas has long anguished?
What forms of citizenship does this new landscape speak to? None of these issues
is predetermined. Each is a subject of economic and political negotiation, playing
out, as we shall see later, within particular ideological formations.

Waterfront redevelopment is wholly a product of our own era, and the past
has been reconfigured to fit the present. The inevitable intertwining of place
and identity – for landscape facts always implicate societal values – means that

the interpretation of the new inner city provides insights on prevalent values in contemporary urban life.

Precedents: The Industrial Port City

The predecessor to current revitalised waterfront settings is the industrial city, the shock city of the 19th century. When travellers visited Manchester in the 1840s or Chicago in the 1890s, or indeed South China's Pearl River Delta in the 1990s, there was the sense of an industrial phenomenon on display, something portentous and without precedent. Rampant industrialisation had minimal checks and balances. It was a libertarian landscape, with, as de Tocqueville observed on a visit to Manchester, 'no trace of the slow continuous action of government' (cited in Marcus 1974: 61). In the pursuit of economic development, social justice and environmentalism were historic sidebars left to marginalised intellectuals and clergymen for commentary. A few years before Friedrich Engels' momentous visit to Manchester in 1842 and his confident conclusion after empirical assessment that 'the revolution *must* come' (1958: 335), de Tocqueville passed through the town in 1835, leaving behind his own celebrated description:

> From this filthy sewer pure gold flows. Here humanity attains its most complete development and its most brutish; here civilization works its miracles, and civilised man is turned back almost into a savage. (Cited in Marcus 1974: 66)

De Tocqueville also noted the importance of water transportation and how the meandering streams and small rivers of the district had been straightened into industry-serving canals by the calculating eye of commerce. At the end of the century would come the Manchester Ship Canal, transforming the land-locked industrial town into a port city. In an age of growing international trade in bulky raw materials and other commodities docklands rapidly expanded; the Port of London added 10 new docks in the 19th century, beginning with the West India Dock in 1802 and ending with Tilbury Docks in 1886. Industry and warehouses crowded the docks and the railways that led to them. The commercial returns on industry were vast and industrial space invaded other land uses. As the historical geographer Peter Goheen (1970: 11) observed in the lakeshore port of Victorian Toronto: 'industry was able to demand almost any land in the city. Such was its bidding power and such was the utility which manufacturing gave to the land'.

Around the port and the manufacturing zone were the cramped and poorly built quarters of the working class. As urban transportation expanded residential options, the middle class and the elite withdrew from these noisy, crowded, polluted and unhealthy sites. The inner city assumed a spoiled identity, consolidated by the influential writing of the Chicago School of urban sociologists in the first half of the 20th century, who saw this urban region to be an inherent source of multiple forms of social disorganisation. Bourgeois homes in the inner city *filtered down*, in the value-laden terms of 1930s land economists, to apartments and rooming houses occupied by successively poorer households. Toronto's 1941 map of housing and land use contained such an innermost zone around the downtown core and adjacent to industrial districts, bluntly described on the map as 'fourth-class housing'.

The term 'fourth class' immediately suggests the basis for improvement, and after 1945 the nascent welfare state, governed by high modernist ambitions, frequently undertook widespread clearance and so-called urban renewal in such districts, not realising that there were circumstances when to be old and crowded could also be the basis for tightly-networked community life. In any case, modern architecture was not enough to right the poverty and social problems of these districts and massive housing projects arguably contributed to a rising alienation and marginalisation of the poor. So over the course of 100 years the inner city was constructed not only as a place but also as an idea. The view of many policy-makers was summarised in Margaret Thatcher's unguarded frustration with what she called 'those inner cities' following her victory in the British general election of 1987 (Robson 1988).

The New Inner City

But in fact a new inner city was already being defined. By the 1960s an artistic scene was gathering especially in large metropolitan areas, Greenwich Village in New York, Haight-Ashbury in San Francisco, Yorkville in Toronto, Chelsea in London. The inner city was cheap for sure, but it was also edgy, sufficiently detached from middle-class society that cultural and political experimentation was possible, while artistic grenades could be lobbed in the direction of the safe and conservative suburbs (Ley 2003). Moreover, besides the growth of an artistic sub-culture, the emergent post-industrial service economy was creating a new army of foot soldiers attentive to the charms of urban living. Jane Jacobs' immensely significant Greenwich Village based narrative, *The Death and Life of Great American Cities* (1961), helped to cause a paradigm shift in the perception of old inner-city neighbourhoods. The charms of urbanity were at first most attractive to a left-liberal new middle class of young professional workers of the 1960s generation who were typically employed in junior positions in health, education or welfare in the vastly expanded welfare state, or else were in the media or design fields (Ley 1996). They were rich in cultural capital like the artists but held more substantial economic capital.

Inner-city housing markets began to move in new and unexpected directions. Informal and self-financing aided home purchase while sweat equity accomplished necessary housing renovations among the early members of the new middle class, a precedent that soon gave way to renovation firms and then niche development companies. In Canada by 1970 the condominium or self-owned apartment was being marketed, offering homeownership equity in the inner city to small middle-class households. The break-up of the rental market was also accompanied by widespread renovation of rental rooms, de-converted into lower density flats, announcing the arrival of filtering up among inner-city housing units as an alternate trajectory to the classic model of filtering down. As the market expanded, banks and other lending institutions joined the bandwagon and available credit facilitated home purchase. Newspapers and magazines advertised new hot lifestyle districts and wealthier households and property speculators moved in. By the 1980s in large Canadian cities one was more likely to find a professional living in the inner city than in the suburbs (Ley 1996). In the 1990s the market expanded to include empty-nest households in their fifties and older, who could afford the growing costs of downtown living and who, with children out of the house, were able to take advantage of the rising arts and leisure opportunities of the downtown. The housing market of the new inner city has expanded far beyond the young urban professional who started all the excitement.

Waterfront Redevelopment: The Case of False Creek, Vancouver

Once the success of smaller projects undertaken by individual households and small builders in the inner city was proven, large developers and, later, international capital arrived. Formerly sceptical of the potential of the inner-city housing market for the middle class, larger corporate players were drawn in part to this profitable niche market by the availability of obsolete industrial, warehouse and transportation sites. Brownfield sites on vacant industrial land offer larger property bundles – and thus more comprehensive planning and development opportunities – that cannot be matched by the single lots or small assemblies in long-standing inner-city residential districts. Many of these brownfield industrial sites were on the waterfront, for bulky commodities were often moved by ships and barges. In addition abundant water sources were used (and abused) in cooling and other manufacturing processes.

The early American prototypes of waterfront redevelopment like Baltimore Harbor involved significant public sector participation, for risk-averse private corporations were as yet unwilling to venture into the untested experiment of converting industrial land to residential, commercial and leisure districts. Vancouver's False Creek, where there have been three distinct phases of waterfront development over a 30-year period, shows dramatically the emergence of waterfront districts as a successful mixed-use landscape, a landscape always inscribed with larger societal ideologies (see Figure 4.1). The changing position of the state in all three public–private initiatives offers a view of changing values and capabilities from the optimistic 1970s to the crisis-ridden present.

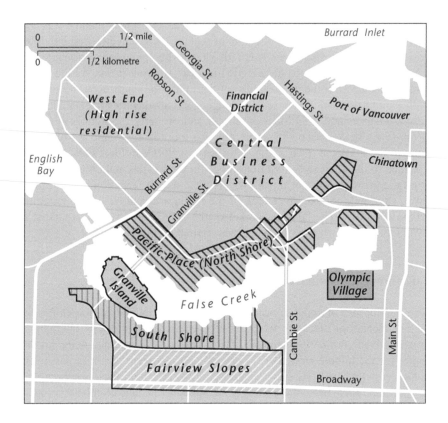

Figure 4.1   False Creek, Vancouver

False Creek is a long ocean inlet that transforms the downtown area into a peninsula, and in earlier times, before extensive land reclamation shrank the water body, into virtually an island. It became the city's major industrial region early in the twentieth century, with a swath of waterfront industries including the transshipment and simple processing of the regional staples of lumber, salmon and minerals, as well as the manufacturing of essential regional products, including tugs and barges, mine- and mill-working equipment, and basic industrial products like chains, barrels and even nails. The whole waterfront was formerly exclusively zoned for industrial uses, was the terminus of several long-distance railway lines, and a celebrated region of industrial pollution. As recently as a 1968 report, False Creek was described as 'a garbage dump, a sewer outlet for the city of Vancouver' (Fukui 1968).

But at the same time an urban reform movement of young professionals discerned a new urban future for the City-owned land on the south shore of the Creek, and as the vision spread, they sought to give shape to an aesthetically pleasing mixed-use environment that would include self-owned and rented apartments, leisure uses, and a specialised retail precinct, Granville Island, built in reconstructed industrial premises. From a negligible residential population, numbers have risen to about 15,000 residents around the shores of False Creek at present, and are planned to reach 25,000–30,000 in a decade. Since the 1970s there have been three distinct phases to redevelopment and they show, as contemporary ideologies shifted, varying commitment to the triple objectives of economic viability, social justice and environmental sustainability. Besides historical contingencies, the three projects also reveal a geographic re-scaling from regional and national to global imperatives.

The first redevelopment, from the mid 1970s to mid 1980s occurred on the under-utilised industrial sites on the south side of the Creek, in projects undertaken by the City and Federal governments. The project reveals a strong 'small is beautiful' mandate; in addition to residential enclaves, the old industrial site of Granville Island was transformed at the same time into a mixed-use festive space (Cybriwsky et al. 1986; Ley 1987). The large expanse on the north side of False Creek was re-shaped under the aegis of a quite different ideology. Old manufacturing plant and railway lands had been consolidated by the provincial government in a land swap, and transition began with the demolition of waterfront industry to make way for the 1986 World's Fair, Expo 86. Two years later the provincial government, enamoured of Margaret Thatcher's privatisation policies, sold the whole site, which was re-named Concord Pacific Place by its new Hong Kong owners. Built out over the period from 1990–2010 Concord Pacific reveals the development impulses of neo-liberal globalisation (Olds 2001). The third waterfront landscape, currently under construction, is Southeast False Creek. On former City land, it is a project that removes the final area of industrial land use from the waterfront. Conceived to include the athletes' village for the 2010 Winter Olympics, Southeast False Creek is heavily impregnated by global marketing but also by the deep sustainability objectives of an environmentally anxious world. Its international exposure has been completed by the failure of its New York financier during the global financial crisis of 2008–2009. These three projects show markedly different landscapes and founding ideologies. In particular I am interested in how they incorporate in different ways economic viability, social justice and environmentalism into a planning and design solution.

False Creek Southside with Granville Island: The Small is Beautiful 1970s

The land on the south shore, False Creek Southside, was owned by the City of Vancouver and had long been leased to industrial users. Following a political reform movement in the late 1960s, a new city council of young liberal professionals declined to renew industrial leases as they ran to term, and proposed in 1972 an innovative mix of residential, recreational and commercial uses to replace declining industry (Cybriwsky et al. 1986; Ley 1980). By the mid 1970s a new landscape was taking shape that on completion a decade later housed almost 6,000 people. It is a template of 1970s liberalism, built at the high watermark of the welfare state, when public finances were still capable of significant social planning and innovation, and neo-liberal impulses still lay in the future. Land uses included a high level of public green space landscaped in a regional idiom, a lake, a paved waterfront seawall for pedestrians and cyclists, and clusters of low-rise housing with varied designs, many clusters arranged like a doughnut around a central open space to facilitate social interaction and community building in the semi-private core. The contrast with the more formal Concord Pacific Place on the opposite north side of False Creek is marked, for these residential towers in public open space, sired by a Hong Kong model of dense urban development, are closer to an older Le Corbusier model of high-rises in an uncluttered open space setting.

Government leadership on the Southside was essential because the private sector was risk-averse in confronting what in the early 1970s was a totally unfamiliar development scenario. No large private financial backers or property developers were willing to participate in building housing on former industrial lands, and the project was launched on government funds and the vision of small local builders. The 1970s ideologies of social justice, a critical postmodernism, and environmentalism are writ large in the built environment. Taking advantage of new social housing programmes to promote an egalitarian objective, *social mixing* was an integral component of the residential programming, and housing was built to be one-third low income, one-third middle income and one-third high income. In the decade in which multiculturalism was launched as a national ideology, social and cultural diversity and varied tenure forms were emphasised. *Critical postmodernism* was evident in the small is beautiful, human-scale design, the diversity of mainly low-rise building forms and materials, the attention paid to visual aesthetics, the scaling of public and private space, and a constrained symbolic programme including iconic rooflines built to simulate the railway wagons that had formerly occupied the site. Design documents were explicit in their rejection of high-rise precedents in earlier private and public renewal.

The *environmentalism* of the 1970s, particularly prevalent in the city where the Greenpeace Foundation was established, was visible in an ambitious green space programme. Public open space was abundant, and other amenities included a lake with a nesting area for Canada geese, a woodland trail through a simulated West Coast forest glade, and an engineered waterfall and stream beside a garden of native species. Physical, social and environmental planning were conjoined in an ideologically progressive landscape that owed much to the political visionaries of a post-industrial city, with a strong welfare state reproducing regional and national ideals in a landscape narrative that has won widespread approval from planners and urban designers. Here is a landscape where internationalism, let alone globalisation, makes only the most meagre appearance, and is trumped by familiar icons of province and nation reproduced in a distinct historic period.

The City's largely residential project was built in conjunction with the recycling of the adjacent old industrial site of Granville Island, owned by the federal government, and joined to False Creek Southside by a short causeway. Granville Island had always been an industrial site, formed from dredging and infill around a sand bar in False Creek during the First World War when there was a premium on land dedicated to industry. With a remarkable touch of the federal magician's wand, this district of dilapidated manufacturing was transformed into a mixed use leisure, office and retail location that includes a successful Public Market, an upscale hotel, marine and professional firms, theatres and arts and craft stores, plus the city's university of art and design. In an unusually principled stand on land use mixing, the planners elected to retain a viable cement-making plant on the island and encourage the continued existence of residual manufacturing. But while the cement plant has remained, old industries have died off through consecutive recessions, although a boutique beer-maker has made a successful landfall.

Granville Island has become one of the City's landmark sites, and was named the 'Best Neighborhood in North America' in 2004 by the New York-based Project for Public Spaces. The celebration of familiar regional icons has been taken even further on Granville Island than through the Southside development, for the dominant industrial vernacular styling has retained and even duplicated rough industrial shells, made from corrugated iron, but now containing cultured post-industrial interiors. Here is what Sharon Zukin (1989: 174) described in New York's Soho as a 'poetic appreciation of industrial design'.

Concord Pacific Place: Selling a Global Waterfront

Meanwhile, and in sharp contrast, the north shore of False Creek, owned by the provincial government, was selected in the early 1980s as the site for the World's Fair, Expo 86. The decision to hold the Fair was made during the deepest recession in half a century and was a clear effort by the government to kick-start the provincial economy through the now familiar strategy of a global leisure spectacle. Cutting back on public services and taking on the robust union movement guaranteed political polarisation and brought the province within a whisker of a general strike called by a left-liberal alliance, the 'Solidarity Coalition'. Expo 86 was part of a neo-liberal exercise in re-imagining Vancouver's global branding by supporting the new service economy of tourism and the cultural economy, as well as by promoting immigration and foreign direct investment, notably in the property market (Ley 2010). In contrast to the inward-looking regional and national project on the south shore, the north shore development has been all about conveying a message of open borders, proclaiming Vancouver and British Columbia's destiny as part of the freewheeling Pacific Rim.

The right-wing provincial government was committed to global free markets (Mitchell 2004), and at the end of the Fair it privatised the Expo site in an international land sale. By this date, 1988, there was no difficulty in attracting big money for waterfront redevelopment in Vancouver, and a conglomerate of wealthy Hong Kong corporate giants, led by Li Ka-shing, the wealthiest capitalist in the colony, bought the entire two-kilometre block of land, and initiated a 20-year build-out of the extensive site (Ley 2010; Olds 2001). The redevelopment model has been very different from the south shore, led by the private sector, but with the City negotiating public amenities like access to a public seawall trail and open space and community services, while requiring that 20% of the land be reserved for subsidised housing, though there was no obligation on the developer to construct the dwelling units. The outcome of the presence of a single large developer, unlike the myriad of small builders on the south side, has been a much more standardised landscape – repeated, though with fewer amenities, in Pacific CityPlace, a later brownfield project in Toronto by the same developer. The appearance is if anything a replication of a Hong Kong skyline with tall condominium towers themed with clear design affinities. While densities are more than twice as high as the south side and there is much less socio-economic diversity, nonetheless the north side is certainly more cosmopolitan with a higher share of foreign-born households and non-citizens. Especially in the early years, there were many speculative offshore sales, while name recognition of the developer led many wealthy immigrants from East Asia to make their Canadian landfall in the Concord Pacific apartment towers.

The project is also much more accessible to the car. The Southside, built just after Vancouver's great freeway conflict that rejected a freeway network for the city, turned its back on the car, marginalising it to peripheral and underground locations, while Concord Pacific Place is more accommodating to the private automobile. The project is penetrated by the continuation of the existing downtown street system and includes on-street parking. Its leitmotif is pragmatic urbanity rather than small-is-beautiful romanticism. Stores are located on these through streets, and the project is bisected by two large sports stadia, for football and ice hockey, which are significant traffic generators.

Concord Pacific Place is a cosmopolitan, corporate and generic landscape, attuned to the globalisation of the 1980s and 1990s rather than the national, welfare state ideology of the 1970s. Its marketing too has been international, with sales offices not only in Vancouver but also in Hong Kong and other East Asian cities. Indeed properties were often sold in Hong Kong and elsewhere in East Asia to investors and potential immigrants after only brief marketing exposure in Canada. The regionalism and nationalism of False Creek South clashes with the global horizons of the waterfront towers on the north side of the Creek. The effervescent and innovative state that shaped the south shore has been visibly downsized and weakened in its capacity and imagination on the north shore.

Southeast False Creek and Global Crisis

Southeast False Creek is the last district to encounter redevelopment, which began in earnest in 2008. Most of the site was again publicly owned (by the City) including the water frontage. Here we see internationalisation taken a stage further, for part of Southeast False Creek (SEFC) is to provide the Olympic Village for the 2010 Winter Games after which apartments will be sold as private condominium units. The Development Plan dates from 2005 and reflects the most current political and planning priorities. The principal thrust of a long planning process is that SEFC should be 'a model sustainable community' (City of Vancouver 2009). With the smallest footprint of the three False Creek developments, SEFC will accommodate between 12,000 and 16,000 people by 2020, with a density of up to 36,000 per square kilometre or five times the False Creek Southside densities.

What is interesting here is the heavy emphasis on environmental objectives in the design and planning process. This is in line with the City of Vancouver's current planning policy goal of 'ecodensity' as a municipal response to global climate change, the argument being that the raising of densities will facilitate walking, cycling and transit, and limit fossil fuel consumption by the private car. It should be noted that the provincial government in British Columbia became the first jurisdiction in North America to legislate a carbon tax in 2008. Of course the first False Creek project, the 1970s Southside development was also, as I have shown, shaped by environmentalism. But this was first wave environmentalism, defined by the inclusion of regional ecosystems and green space. Now environmentalism is more ambitious. It is re-scaled to a crisis at the global level, and is extended to the details of building design. The LEED gold standard certification at the neighbourhood scale in SEFC includes rainwater recycling, green roofs and a sustainable energy system; besides solar power, sewage sources will be used for heating the project. Space is reserved for urban agriculture and an ecological reserve. Shops and services are to be within walking and cycling distance, while bus and rapid transit will facilitate off-site movement. SEFC won the Best Overall Project award for brownfield redevelopment from the Canadian Urban Institute in 2009.

The existing development plan was, however, narrowly passed by the City Council after protracted disagreement. Conflict emerged over the place of social housing in the project. The long planning process had ambitiously sought to include social as well as environmental sustainability, but eventually a right-wing Council threw out all but a small portion of affordable units, limited at present to 250 out of a stock of at least 6,000 units. One of four guiding principles insisted on by the Council was 'viability without subsidy' (City of Vancouver 2009). Unlike the Southside development 30 years earlier there will be a single private sector developer, though the City will manage the planning and design process on leasehold land. Although building began in 2008 and the first apartments will not be available for occupancy until after the Olympics in 2010, marketing of units began in October 2007. The cheapest apartment units in Phase 1, with less than 600 square feet (55 square metres) are priced at CAD$450,000 (300,000 euros). Prices go from there up to $3.5 million.

Despite its density, the Olympic Village is too expensive for poor people. Earlier plans for significant social housing have been reduced to 250 units, with the recognition that these units will cost $595,000 each to build and require continuing monthly subsidies of perhaps $2,000 per unit to bring mortgage repayment to an affordable level (Cernetig 2009a). These are unsustainable costs in an era of substantial public debt and more ambivalent commitment to the public household model of the Canadian welfare state. There are profound anxieties among housing advocates that ecodensity will aggravate housing affordability and be another force priming the rising inequality that runs the poor out of Canadian inner cities.

The global parameters of SEFC reach deeper than environmentalism alone. The project has been financed from New York by Fortress Investment Group, a Wall Street hedge fund. In the global financial crisis in 2008, the company's stock prices collapsed, and the City of Vancouver was forced to underwrite construction financing with a loan of $100 million to the Fortress Group and another $190 million in loan guarantees (Cernetig 2008). As the financial crisis deepened, Fortress requested more substantial loan guarantees from the City, amounting to 'most' of the $750 million financing for the project (Cernetig 2009b). In the worst-case scenario of a public–private partnership, the City of Vancouver has become the lender of last resort, shouldering the principal risks of project viability in an unpredictable real estate market. Cost overruns have occurred, the project is not on schedule, and the condominium market for a high-priced property is less buoyant than it was when construction began, so that the marketing timetable has been extended three years beyond its original schedule. With a re-financing package in place from Canadian banks, a KPMG audit showed that the City still remains the ultimate guarantor for close to $1 billion (Cernetig 2009c).

As an Olympic site, Southeast False Creek is a project with a more complete global leisure and marketing prospectus than Expo 86. Planned in response to a global environmental crisis, it has become caught up in a global financial crisis. The local state has become the caretaker, cleaning up the mess once the party is over. The project at present looks like a prime example of life in Ulrich Beck's risk society.

Conclusion

False Creek shows several diverse models of waterfront redevelopment that have been undertaken over a 30-year period, each of them reflecting and reproducing certain dominant interests and values in the society of the day. The original Southside development with its firm roots in social justice, environmentalism and a critical postmodernism evokes very much a time and place, Canada in the Trudeau era of the 1970s, with an innovative welfare state leading the way while the private sector looks on nervously. Economic returns were not a goal, as the City sought to break even on its land. Twenty years later, much had changed. We see instead a model of neo-liberal globalisation on the Expo site, following the Pacific Rim marketing of a World's Fair. Large international developers confidently entered the picture, creating generic urban spaces and reaching a pan-Pacific market. But with the democracy of the market, prices were moderate, and some public benefits were negotiated; environmentalism was not a feature of this development other than initial soil remediation. Currently, in Southeast False Creek, the global environmental crisis drives the design process while an international developer sees the market opportunities, in what his website calls 'a world address'. Millennium Water, to use the marketing name, is scrupulous in its pursuit of environmental sustainability in response to a global crisis, but at what cost? Even on city-owned land, all but wealthy households are ruled out of a landscape shaped by the City of Vancouver's ambivalent policy of ecodensity. But the upscaling of the project introduces new vulnerabilities to the local state. The global financial crisis led by the collapse of Wall Street in 2008 imperils the project, and the City of Vancouver is left in the precarious state of final guarantor to ensure that the site is ready for the 2010 Winter Olympics. Rooted in space, the local state cannot evade its global responsibilities.

References

Cernetig, M. 2008. An Olympic Veil of Secrecy. *The Vancouver Sun*, 7 November, A1.

Cernetig, M. 2009a. Olympic Village: Social Housing to Cost $595,000 per Unit. *The Vancouver Sun*, 23 February, A1.

Cernetig, M. 2009b. Olympic Village Shock. *The Vancouver Sun*, 9 January, A1.

Cernetig, M. 2009c. Olympic Village: Plagued by Cost Overruns and Construction Delays. *The Vancouver Sun*, 7 October, A1.

City of Vancouver 2009. *Creating a Sustainable Community: Southeast False Creek* [Online]. Available at: http://vancouver.ca/commsvcs/southeast/index.htm [accessed: 2 November 2009].

Cybriwsky, R., Ley, D. and Western, J. 1986. The Political and Social Construction of Revitalized Neighborhoods: Society Hill, Philadelphia, and False Creek Vancouver, in *Gentrification of the City*, edited by N. Smith and P. Williams. London and Boston: Allen & Unwin, 92–120.

Engels, F. 1958. *The Condition of the Working Class in England*, edited and translated by W. Henderson and W. Chaloner. Oxford: Blackwell.

Fukui, J. 1968. *A Background Report on False Creek for the Vancouver Board of Trade*. Vancouver: Board of Trade.

Goheen, P. 1970. *Victorian Toronto: Pattern and Process of Growth*. Chicago: Department of Geography Monograph Series, University of Chicago.

Jacobs, J. 1961. *The Death and Life of Great American Cities*. New York: Random House.

Ley, D. 1980. Liberal Ideology and the Post-industrial City. *Annals of the Association of American Geographers* 70(2): 238–58.

Ley, D. 1987. Styles of the Times: Liberal and Neoconservative Landscapes in Inner Vancouver, 1968–1986. *Journal of Historical Geography* 14(1): 40–56.

Ley, D. 1996. *The New Middle Class and the Remaking of the Central City*. Oxford: Oxford University Press.

Ley, D. 2003. Artists, Aestheticisation, and the Field of Gentrification. *Urban Studies* 40(12): 2525–42.

Ley, D. 2010. *Millionaire Migrants: Trans-Pacific Life Lines*. Oxford: Blackwell-Wiley.

Marcus, S. 1974. *Engels, Manchester, and the Working Class*. New York: Random House.

Mitchell, K. 2004. *Crossing the Neoliberal Line: Pacific Rim Migration and the Metropolis*. Philadelphia: Temple University Press.

Olds, K. 2001. *Globalization and Urban Change: Capital, Culture and Pacific Rim Mega-Projects*. Oxford: Oxford University Press.

Robson, B. 1988. *Those Inner Cities: Reconciling the Economic and Social Aims of Urban Policy*. Oxford: Clarendon Press.

Zukin, S. 1989. *Loft Living*. New Brunswick: Rutgers University Press.

Chapter 5

Planning for Creativity:
The Transformation of the
Amsterdam Eastern Docklands

Robert Kloosterman

[B]y the 1970s the shipping companies abandoned the area to artists and squatters. A decade later architects were commissioned to spearhead a complete regeneration of the docklands. Functional modern housing that championed a slick, contemporary aesthetic attracted well-heeled creatives and young families and a new community was born.

Bergmans (2007: 11)

From Industrial to Post-industrial Uses

The outlook of Amsterdam seemed bleak at the end of the 1980s. The city had suffered from a decline in population as many middle-class families left for the greener pastures in the suburban towns surrounding Amsterdam. The city's economy, moreover, also had suffered heavily from deindustrialisation and the departure of large-scale wholesale and logistical activities. As a result of this often large empty spaces emerged where once factories, docks and marshalling yards were located. These derelict spaces were located close to the city centre and in some cases became home to a more or less marginal urban population of artists, squatters and homeless living in old warehouses, caravans, tents and huts as happened to the Eastern Docklands (Blokker 1995; Mak 1992: 236).

These developments were, of course, not confined to Amsterdam. Other cities suffered from the same fate of a combination of an erosion of the employment base and an outflow of middle-class households. Agglomeration economies appeared to be surpassed by agglomeration diseconomies as congestion, lack of space, unsafe streets and air pollution. If the glue of agglomeration economies becomes unstuck, the urban fabric, inevitably, unravels.

Urban environments, however, proved much more resilient and resourceful and they became attractive again for work, living and leisure and cities bounced back in the closing decades of the 20th century (Beauregard 2002; Cheshire 2006; Scott 2008). Especially capital cities, such as London, Paris, but also much smaller Amsterdam, experienced an urban renaissance. According to Peter Clark (2009: 242) these capitals 'demonstrated a striking ability to recover, aided by their dominance of national and international business and transport networks. The upturn in world banking and finance since the 1980s consolidated their primacy'. With the wisdom of hindsight, we can see that already in the second half of the 1980s, when observers were still contemplating the imminent demise of large cities as viable places, things had taken a turn for the better. First, employment rebounded as transaction-intensive and, therefore, mostly urban-oriented activities (initially especially various producer services) expanded. The deepening of the process of globalisation with value chains covering ever more disparate locations contributed to the growth of spatially concentrated control and co-ordination functions. These transaction- and contact-intensive functions, together with an elaborate supportive infrastructure of lawyers, consultancies, financial advisers, etc., turned out to be drivers of strong economic growth in large cities (Sassen 1991). Second, cities also re-emerged as attractive places for urban living, notably for large numbers of young, highly educated persons (Butler 2007; Ley 1996). Their penchant for more individualised and, arguably, more reflexive lifestyles kept them in the cities where they graduated or they moved to other cities where they could find the jobs and the amenities to support and develop a certain lifestyle. The British writer Jonathan Raban was already clearly aware of this in his farsighted book *Soft City*, which was published in 1974. He depicted London as a complex mosaic of streets, squares, (cheap) apartments, shops, cafés, restaurants and exciting public spaces. Together they created a huge emporium with an extremely wide range of products, images and ideas – from the mundane to the bizarre – with which young, upwardly mobile individuals could construct and assemble their own lifestyles and identities.[1]

Whereas the rise of producer services after 1985 made its mark on urban landscapes chiefly in the form of concentrations of high-rise office towers, the 'new urbanites' showed up as initiators of processes of gentrification. Lots of former 19th and early 20th century working-class neighbourhoods were converted into vibrant (and eventually expensive) urban areas with telltale amenities such as vegetarian restaurants, galleries, design shops and boutiques (Zukin 1995). The gentrification did not remain limited to just residential areas from the industrial era. One can also witness the transformation of large-scale central urban industrial workplaces – such as factories, warehouses, beer breweries, gasworks,

---

1   Almost 25 years later, Jonathan Raban described London in 2008 in the *Financial Times* (9 August). The soft city he knew had turned into a much harder city which was more difficult to access due to the very strong rises in rents even in neighbourhoods that used to be relatively cheap.

railroad yards and docklands – into apartments, office spaces, shops, restaurants, cafés, theatres, museums and other exhibition rooms. One can even argue that the marginal urban population that first populated these abandoned spaces in the early 1980s was a harbinger of a new era to come with very different uses as happened in New York when that city suffered from a deep crisis and artists moved into empty lofts (Currid 2007).

The set of changes in function after 1985 exemplify the transition from an industrial use of urban spaces to a post-industrial one and this process is intricately linked to the underlying shift in the technological mode of production from mechanical to digital. The cultural industries – now one of the mainstays of the digital city (Scott 2007) – figure prominently not just in bottom-up processes of gentrification with artists moving into deprived neighbourhoods, but also play an important role in top-down urban regeneration with cultural amenities spearheading the renewal of derelict urban landscapes as exemplified by the Guggenheim Museum in Bilbao (Gomez 1998). Below, we will examine the role of cultural industries in two cases of urban regeneration; both situated in the Amsterdam Eastern Docklands and then assess the part played by local policies. The first case presents an attempt to create a cultural district with large-scale cultural amenities close to the Amsterdam Centraal Station. The second case consists of another part of the Amsterdam Docklands, somewhat more to the east, which has been designated as a residential area and where, unintended, an increasing number of home-based businesses in the cultural industries are surfacing (see Figure 5.1). By looking at the transformation processes of two adjacent areas, each located in the Eastern Docklands, a former large-scale industrial/logistical area, we will be able to shed light on the impact of urban planning within the context of the emerging digital urban economy. The process of transformation of urban areas does not occur in a vacuum, but instead has to be situated in a broader socio-economic context. Before presenting the two cases, we briefly sketch the broader context of urban regeneration plans, by a more general assessment of the socio-economic characteristics of the digital city.

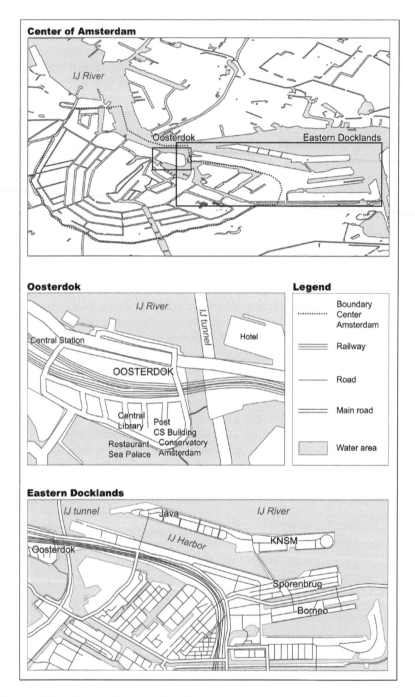

Figure 5.1   Amsterdam docklands

A Closer Look at the Digital City

Urban living has been around for more than seven millennia (McNeill and McNeill 2003). Cities have struggled, persisted and often thrived in pre-industrial, industrial and, now, in post-industrial times. Cities have used their internal agglomeration economies – based on the classic urban triad of proximity, diversity and (critical) mass – as well as their external linkages which enabled them to specialise in certain forms of production and consumption (Hohenberg and Hollen Lees 1995). Technological change and shifts in external linkages have continuously shaped and transformed urban landscapes. The most recent wave of structural changes has occurred in the last quarter of the 20th century when the introduction of digital technologies in conjunction with a rapidly expanding (global) network of external linkages has fundamentally reshaped the economy of many cities. Digitalisation and globalisation enabled cities to return, after an industrial intermezzo of about two centuries, to their essential functions: 'l'échange, l'information, la vie culturelle et le pouvoir' according to the French historian Jacques Le Goff (1997). More precisely, advanced consumer and producer services (l'échange, l'information), co-ordination of spatially disparate tasks in the private and the public sector (le pouvoir) and cultural industries (la vie culturelle) have led the way in creating vibrant economies once again in especially capital cities with already strong historical roots in these activities (Clark 2009; Currid 2007; Currid and Connolly 2008; Hall 1998; Kloosterman 2004; Pratt 1997; Scott 2004: 461). This apparent return to age-old urban functions involves, however, more than just a changing of the guards (Evans 2004). With the advent of a new production paradigm, not just dominant sectors change, but, moreover, the way production takes place and what is produced alter (Hutton 2004; Scott 2008). Use of spaces, from the scale of an individual dwelling, that of the street, the neighbourhood, and that of the city itself are affected by these changes in ways of production and consumption. New spatio-temporal patterns of work, care and leisure are now emerging in advanced cities which are fundamentally different from the dominant patterns in the industrial age with sharp division in space and in time between work and other activities (Florida 2002). Any attempt to describe these fundamental changes will fall short of describing the complex reality of cities. To get a handle on the more general changes, we have summarised them under four headings.

Leading Sectors

The combined – and interrelated – forces of digitalisation and globalisation have given rise to a post-industrial urban profile (Hall 1999; Scott 2008). The first component of this profile is anything but new. Cities are still the main sites of power and influence both in the public realm (the public sector remains a pillar of urban economies) and in the private sector (headquarters of multinationals).

The second component is related to these monitoring, control and co-ordinating activities, namely the advanced producer services which have emerged as a strong engine of urban growth after 1985. These services – from management consultancy to financial advice and law firms – have been pivotal in supporting the increasingly spatially fragmented chains of production and consumption (Sassen 1991). They are strongly dependent on external economies (specialised labour pool, amenities, spillover) generated by proximity and are, hence, mainly concentrated in (large) cities (Hall and Pain 2006). The current financial crisis may put an end to the spectacular growth rates of the recent past, but these services will remain important in organising the globe-spanning arenas of production and consumption.

Whereas producer services are very much part of the private sector, the third component, the cluster of higher education/healthcare/research, is in many cases closely linked to, if not part of, the public sector. Universities, university hospitals, research institutes and higher vocational educational institutions are evidently important as generators of urban employment growth. The opening up of higher education to the middle and even working classes after the Second World War has helped to turn centres of higher education into urban growth engines (Hobsbawm 1996; Judt 2005). The importance of knowledge as a source of competitiveness in a global economy has more recently once again boosted higher education and research. Healthcare, in most cases institutionally linked to higher educational institutions, has also strengthened urban economies as it is in increasing demand due to high income-elasticity and a greying population.

The specialised consumer services (including tourism) constitute the fourth component of the economic profile of the digital city (cf. Zukin and Maguire 2004). The fragmentation of consumer markets has contributed to the rise of small niche markets catering to specific tastes. Larger cities have the necessary critical mass to sustain these niches. More and more, larger cities distinguish themselves through high-end consumer services with high-quality shopping facilities (e.g. flagship stores), high-quality leisure (e.g. top-class restaurants; nightclubs) and sports facilities both for spectators (e.g. sports stadiums) and participants (e.g. fitness centres). These amenities are not just for the inhabitants of a city, but also for its workers and, increasingly, aimed at visitors and tourists. They evidently contribute to the quality of a place and, hence, following Richard Florida (2002), in helping to attracting high-skilled workers and boost economic development (cf. Trip 2007).

The fifth component comprises those economic activities concerned with producing and marketing goods and services permeated with aesthetic attributes (Scott 2000). Increasingly, aesthetic qualities (material or imaginary through image-building) are used to give products a competitive edge in ever more fragmented and volatile consumer markets (Caves 2000). Cultural industries, from designers to advertising agencies, are in the business of endowing products with these aesthetic qualities. The products of these cultural industries are either immobile as they too have to be consumed at the place of production (Scott 2006), as in the case of theatres, museums and festivals, and here they partly overlap with the high-end consumer services. Or the products are mobile in which case they can be, in principle, consumed anywhere: recorded music, design products, books, fashion, etc. Cultural industries are, in general, dependent on the classic triad of Marshallian agglomeration economies: specialised labour pool, a dedicated infrastructure and a more or less continuous spillover of knowledge (Kloosterman 2008). This makes them strongly oriented towards cities and, within cities, towards city centres. A high-income elasticity has pushed up demand for products with aesthetic attributes and this, in turn, has fostered the growth of cultural industries.

We can find all kinds of hybridisation as, for example, between high-end consumer services and cultural industries. The clusters of economic activities are also interrelated in many ways as high-paid workers in the producer services generate demand for high-end consumer services and cultural industries. The presence of higher educational and research institutions attract a highly educated population which help to sustain local cultural industries and they help to supply the other clusters with highly specialised workers. These activities share a strong urban orientation as they are all dependent – although in different ways – on agglomeration economies generated by proximity, diversity and critical mass. The larger the city is, the higher are, in principle, the agglomeration economies. Moreover, the complex interlocking of the different clusters generates important emergent effects such as localised dedicated institutions and a local culture or atmosphere which partly feeds into its products (such as fashion from Paris, music from Manchester, films from Hollywood or architectural design from Rotterdam), makes it hard to copy and, therefore, gives to some extent protection to competition from elsewhere.

Technological-Organisational Aspects

The arrival of a new production paradigm in the cities did not just involve appearance of new leading sectors. Ways of production have also changed with the advent of cheap computing power in combination with easy access to global digital networks. It has put an end to the dominance of the traditional industrial/ Fordist business model based on high fixed costs and, hence, economies of scale. Small-scale production, supported by a whole array of innovative producer services such as parcel delivery and ICT services (Friedman 2005), has become profitable. Thanks to combinations of flexible production and digital linkages, small firms are able to exploit economies of scope. They cater to often very volatile, niche markets – the long tail of the demand side – that clearly transcend the local. The small firms are usually transaction-intensive, working in local and, increasingly, non-local networks which may underpin complex chains of production. They are typically dependent on central urban locations, which offer the possibilities for relatively cheap face-to-face interactions. We are seeing the emergence of a few large firms – typically focused on marketing and distribution where economies of scale are still important – surrounded by a cloud of small firms delivering various kinds of highly specialised services. The re-emergence of small firms, notably in producer services, consumer services and cultural industries which rely on external economies for their profitability, also reshapes the urban economy.

Labour Practices

With these shifts, knowledge has emerged as the key input for competitive economic activities. Highly refined, specialised and detailed knowledge of products, production processes and markets is key to compete on global but fragmented markets. This has further strengthened the agglomeration forces as (tacit) knowledge, in marked contrast to information which is highly mobile or 'leaky', tends to be more or less place-bound or 'sticky'. The digital revolution, then, on the one hand, increases degrees of freedom with respect to when and where to work and enables people to plug in their laptop at home, in cafés, playgrounds and in parks. On the other hand, as social networks become more important, the ability to meet relevant others in person sets limits to the freedom of choice of the location.

Knowledge-intensive production in combination with digital technology asks for different forms of management. The large-scale, top-down, bureaucratic, Taylorist ways of production are giving way to small-scale, cottage-like production with a strong reliance on self-management. Workers know their targets and they are to some extent free to decide when and where they should work (as in the pre-industrial putting-out system). The rise of self-management can be seen in the rise of (highly educated) self-employed (one-person businesses), but also those who are formally workers are becoming more entrepreneurial. The boundary between being employed and self-employed can be very vague and blurred, especially in the cultural industries with its myriad of small firms and complex combinations of self-employment and temporary, part-time jobs. Increasingly, workers hold a portfolio of income-generating activities which may encompass regular employment (albeit part-time) and self-employment. The traditional (predominantly male) industrial working class has largely disappeared from advanced urban economies in the developed world and been replaced by a much more diverse labour force comprising male and female workers in more balanced way. Self-management does not necessarily imply an improvement. If the workload and insecurity are high, self-management may not be a very attractive option. Whatever the benefits or drawbacks, it is clear that the highly educated workers (Florida's 'creative class') have more degrees of freedom with respect to the temporal and spatial aspects of organising their own work. New combinations of work, care (social reproduction) and leisure arise. Cities, with their array of diverse amenities close by, are attractive places for these new combinations.

Tacit knowledge is hard to transfer over distances as it has to be exchanged in direct face-to-face contacts. Reproduction of competitiveness necessitates (formal or informal) institutions where transfer of tacit knowledge can take place. The clustering or spatial concentration of related economic activities enables the emergence of such institutions where newcomers can learn the tricks of the trade through on-the-job training. In some sectors, as in architectural design, we can find guild-like institutions with highly asymmetrical master–apprentice relationships (Kloosterman 2008). Such localised institutions for the transfer of knowledge further add to the agglomeration economies.

Spatial Consequences

The emergence of a new economic profile in cities, digital technology, new ways of organising production on a firm level and on an individual level are inevitably also articulated spatially. Although the built environment may display significant inertia with respect to changes and earlier historical layers can usually still be seen, the developments described above are evidently altering the use of space in cities. Knowledge-intensive activities, dependent on frequent face-to-face contacts, are concentrated in dense areas of large cities. Such local clusters of specialised activities are pivotal in the global competition. Within these knowledge-intensive activities we should distinguish between those that are able to pay high rents (generally the activities related to power and coordination and producer services) and those that can only afford low rents (mostly cultural industries). The former can be found in prestigious central-business type of locations, while the latter usually opts for transition neighbourhoods (e.g. previous working-class neighbourhoods), quite centrally located and able to sustain frequent face-to-face exchanges but much cheaper.[2] Specialised consumer services can be found in or in the vicinity of both types of locations depending on the niche market they wish to target. Activities less reliant on frequent face-to-face contact are pushed outside and the same can be said for large-scale amenities catering to a mass audience (e.g. sport stadiums).

These changes can be easily observed in many cities. There is, however, another set of changes going on partly behind the façade of the built environment. The shifts in labour practices allow for much more freedom with respect to the temporal and spatial combinations of work. This implies new patterns of mixing work, life and leisure on the level of a city, of a neighbourhood, a street and even on the level of an individual dwelling. The process of an increasing separation of functions, so characteristic of the industrial age, is now partly being reversed. The digital technology in combination with self-management now makes erratic and quite idiosyncratic spatio-temporal patterns of work possible. The digital nomads can be seen working in as people work at home, in cafés, restaurants, museums or other 'third spaces' (*The Economist* 2008). The digital city will experience new uses of urban space and new needs for certain (semi-)public urban amenities.

These, admittedly very general, characteristics should be interpreted as ideal-typical. In addition, one should not fall into the trap of 'epochalist' readings of

---

2   'These inner city New Economy sites and districts comprise not just isolated firms or "outliers", but rather quite substantial ensembles of dynamic industries' (Hutton 2004: 90).

(recent) history whereby one period is contrasted with another, suggesting a complete overhaul. The observed changes do not occur in a vacuum, but instead interact with the concrete local socio-economic, cultural, institutional and morphological context of cities. New layers are added and some old layers are eroded and in this complex process cities will display their own typical, path-dependent, post-industrial trajectories. These more unique traits are the result of the interplay between differences in national institutional frameworks and historical patterns of urbanisation and more local circumstances as the role and position in the (global) urban network and the more long-term socio-cultural and economic orientation of a city (Hohenberg and Hollen Lees 1995; Kloosterman and Lambregts 2007). It is the contingent mix of the local historical context and the more general trends of change which help to shape the concrete opportunities and constraints for individual cities (Hutton 2004). Against the backdrop of these structural changes and its own legacy, a city government has to devise policies to boost competitiveness and attractiveness. More and more, cities have turned to cultural industries as an instrument to foster economic growth, employment and quality of place. 'Heavily backed by national governments, metropolitan authorities have marketed their attractions as businesses, cultural, and leisure centres in often aggressive ways, including redevelopment projects (Dublin's Temple Bar district), cultural grand projects (the Tate Modern, the Gare d'Orsay Museum), and international cultural and sporting events' according to Peter Clark (2009: 243). Before we turn to two cases of the role of cultural industries in Amsterdam, we first take a look at them from a more general point of view.

Cultural Industries as an Instrument of Urban Revitalisation

Cultural industries encompass a wide variety of activities; all geared towards adding semiotic and aesthetic value to goods or services, ranging from music making to advertising and from jewellery making to the staging of theatre plays (Pratt 1997; Scott 2004). Already in 2000, Peter Hall observed that:

> Cities across Europe ... have become taken with the idea that cultural or creative industries (a term that 20 years ago might even have been thought offensive) may provide the basis of urban regeneration. Culture is now seen as the magic substitute for all the lost factories and warehouses, and as a device that will create a new urban image, making the city more attractive to mobile capital and mobile professional workers. (Hall 2000: 640)

The Guggenheim in Bilbao is, of course the prime example of a policy based on the cultural industries, but as many articles in newspapers and magazines testify, it has become a quite common strategy to revamp rundown areas and put cities on the map (again). Hamburg HafenCity with its new concert hall designed by Herzog & De Meuron, but also Sheffield building on its tradition in popular music from the Human League to Richard Hawley and the Arctic Monkeys, are just two other examples in a long line of cities that use culture as an instrument of revitalisation. To bring some order to the various ways in which local culture can be used, the Italian urbanist Walter Santagata (2002) has suggested a fourfold division of clusters of cultural industries each with their own spatial articulation and governance characteristics.

The Industrial Cultural Districts

These are the archetypical clusters of cultural industries with many small and medium size industries specialising in high-value added products and which are embedded in a local production milieu with its own dedicated supporting infrastructure for reproducing the skilled labour, matching demand and supply and maintaining quality standards. The film industry in Hollywood (Scott 2005), the fashion industry in New York (Rantisi 2004) and architectural design in Rotterdam (Kloosterman 2008) are specimens of these industrial cultural districts. These industrial cultural districts originate 'spontaneously': for more or less contingent reasons one firm or a set of firms initially locate in one place, become successful and a process of increasing returns is set in motion which strengthens their competitive position as external economies are generated on an ever larger scale. It would be very hard for a local government to initiate such a cluster, but it could help to accommodate a cluster and assist in creating these external economies (e.g. by supporting dedicated local educational institutions) once it is already there.

The Institutional Cultural Districts

In this case, the competitive position of a particular industrial cultural district is supported by an institutional construction in which the government (national or even supranational) confers the right to the local producers to protect their product through the setting of a collective trademark and the allocation of property rights to a restricted area of production. The particular niche covered by the spatially concentrated producers is then fenced off by formal and legal boundaries further emphasising the importance of monopolistic competition in cultural industries. These districts tend to be more oriented towards the rural concentrations and can be found especially in wine and cheese making. We can find examples of such *institutional cultural districts* (the Piemont-Langhe and Tuscany-Chianti in Italy and the 'Appellation d'Origine Contrôlée' areas in France). If there is a product with strong identity that is linked to a particular locality, then it might make sense for a government to underpin its position by conferring property rights.

The Museum Cultural Districts

These cultural districts are typically found in historical city centres. The anchor of such a district is a museum network often in conjunction with an artistic community. A large part of the production that takes place in these districts is 'immobile' in the sense that it has to be consumed at the place of production: one has to go to the museum or theatre in order to enjoy its products. These cultural anchors are usually part of areas with a wider variety of amenities. The cultural anchors, though, are the main attractors for tourists and visitors. Large cities tend to have several of these *museum cultural districts* within their boundaries, whereas smaller cities typically have just one. These cultural anchors can be historical (Louvre, Paris) or newly established museums (Tate Modern, London). Recently, governments have been using these cultural anchors to transform derelict urban areas as with the Guggenheim in Bilbao or the La Bibliothèque nationale de France instigated by President François Mitterand at the eastern border of Paris. Museums, public libraries and other large public amenities are becoming very important as attractors and as markers of identity for specific locations within cities and for these cities themselves (Rykwert 2002). Given the lumpiness of cultural anchors and the dependence (at least in most European countries) on the public sector for the establishment of a new museum, local governments do have, in principle, a lever here to kick-start a process of transformation of a derelict area.

Metropolitan Cultural Districts

This type of cultural district is by definition part and parcel of the metropolitan environment. It combines aspects of the first type, the industrial cultural, with those of the third type, the museum cultural district. In a metropolitan cultural district we can find, therefore, both the production of immobile cultural products (museum, libraries, theatres, galleries, art shops) and that of mobile cultural products (workshops, studios, ateliers). In addition, the metropolitan cultural district also comprises a supporting structure of amenities (restaurants, cafés, bookshops, magazine stands, nightclubs, 'third spaces' in between work and home where face-to-face contacts can take place). The combination of these different elements (encompassing working, living, leisure) makes metropolitan cultural districts more difficult to initiate than the more mono-functional museum cultural districts. These mixed-use districts tend to be very lively and attractive places to live, work and visit. They can help 'to contrast economic industrial decline, and to design a new image of the city' (Santagata 2002: 16). According to Santagata (2002: 17) two preliminary institutional prerequisites should be met in order to be able to foster the transformation of an area into a metropolitan cultural district: first, the property rights structure should not be too dispersed so economic benefits can be more easily internalised and, secondly, the planning procedure, the marketing and the management of the area should be facilitated by a central body (agency, trust, business community) to handle the complex governance structure. The most important ingredient, namely a critical mass of innovative workers in the cultural industries, is, however, overlooked by Santagata.

In Amsterdam, we can find two examples of emerging cultural districts in the former docklands east of the central station. One closely resembles the museum cultural district while the other one is more difficult to label as its emergence is not so much because of government intervention, but in spite of it.

The Amsterdam Eastern Docklands

When KNSM, the last of the city's shipping giants, tanked in 1977, Havens Oost, the Eastern Docklands area, was earmarked for radical reinvention. Four peninsulas were added to the River IJ and a whole new neighbourhood was dreamed up. Unlike most dreary docklands schemes, though, this one is aimed at aesthetes, not at bankers, and involved the participation of almost every architect in town – around 60 of them at the final count. The result is a modern interpretation of classic tract housing, one of the buzziest neighbourhoods in the Benelux, and, if nothing else, a welcome change of pace from Amsterdam's familiar canalside properties (*Wallpaper* City Guide* 2006: 64).

As in other port cities, the Amsterdam docklands relatively close to the centre were abandoned in the 1960s and 1970s. The increasing size of ships and the containerisation made many docklands dating from the 19th and early 20th centuries obsolete. Large plots of land became available for other uses. Baltimore, Boston and London were frontrunners in redesigning and transforming waterfront and dockland areas into attractive urban areas. In 1983, the city of Amsterdam set up a committee to look at the transformation of the docklands in Amsterdam, the 'IJ-oevers' (banks of the River IJ). The committee concluded that the eastern IJ-oevers indeed had potential but that the area was not very well connected to the rest of the city. The infrastructural links (mainly railways) were connected at the eastern side of the 'finger wharves' (artificial peninsulas constructed to facilitate the transfer of goods from ships to trains and vice versa) and not on the side of the centre of the city. The Eastern Docklands were, notwithstanding their proximity to the centre of the city, quite isolated. A first aim, then, was to open up the docklands to the rest of the city (Bakker 2004).

In 1989, a report commissioned by the city of Amsterdam concluded that parts of the IJ-oevers offered a promising site for large-scale cultural amenities, notably a new music hall. In the early 1990s, the national government, financial partner in the development, demanded that the city of Amsterdam used a public–private partnership to develop the IJ-oevers. The Amsterdam Waterfront Financieringsmaatschappij was established as the vehicle to develop the IJ-oevers. An ambitious master plan for the whole area was drawn up by Rem Koolhaas' Office for Metropolitan Architecture. This master plan was, however, rejected in 1994. The financial base did not seem that sound and, in addition, the plan was seen as too much oriented towards banks and other large financial institutions, too much relying on cars and also as too destructive with respect to the old warehouses that were still there (Lebesque 2006). In addition, the intended occupants of the planned high-rise offices – banks and other financial services – were not inclined to move to the banks of the IJ (Kurpershoek 2007: 395). Instead, real-estate developers had set their eyes on the south-side of the city (the Zuidas), much more accessible by car and closer to Schiphol.

The city had to change tack and a new strategy was devised. This time, the plans for developing the area were much less ambitious in the sense that the existing situation was much more taken as a point of departure (including the warehouses which were to be restored) and, moreover, instead of one plan for the whole area, the area was subdivided into several plots each with its own plans for development and time horizon.

This view was anchored in the IJ memorandum which was accepted by the city council in 1995. This memorandum was:

> conceived in terms of the process for conquering the area. The basic idea was: chop the South Bank up into pieces, and take over the archipelago concept formulated by Koolhaas. Each island is given an identity of its own, and as many of the existing buildings as possible are preserved. The relation with the existing city is brought about by the deliberate placing of a number of 'Anchors': public attractions such as a building or a square. Breaking the plan down into sections enables development per area, while the water and infrastructure guarantees the links. An important point was that this separate development of each area was independent of the structure. A separate theme could be formulated for each island, and several developers would work on it. (Lebesque 2006: 15)

The Oosterdokseiland, the part of the Eastern Docklands closest to the central station, was to become a cultural anchor, whereas the peninsulas further to the east – Java, Borneo, Sporenburg and KNSM islands – were to become mainly residential areas. Below, we will very briefly analyse their developments by using the perspective described above.

Case 1: The Planning of a Cultural Anchor

The Oosterdokseiland or ODE was conceived in the 1990s as a new cultural anchor for the city of Amsterdam. In retrospect, one can label the plans for this part of the IJ-oevers as aiming at a museum cultural district. The city government was strongly involved in planning which amenities should be located there. The new city library, a new music hall and the new building for the Amsterdam Conservatory were all part of the plan. The plan also included parking places, office space, restaurants, shops and housing to achieve a mix of uses at the highly central location (a walk of only five minutes from the central station). A large part of ODE has now been realised. The music hall (designed by 3xNielsen), the library (designed by Jo Coenen), the Conservatory (designed by Frits van Dongen of the Architecten Cie) are now completed and the same can be said for the nearby Cruiser Terminal Hall. The apartments, shops and restaurants will be completed in the second half of 2011. The qualities in the long run of the ODE complex cannot be assessed, evidently, at this point in time.

We can, however, point at some interesting elements in the planning and even already in the implementation of this cultural anchor.

The Oosterdokseiland is clearly a top-down project in which the local government – the city of Amsterdam – in collaboration with the real-estate developer creates a cultural anchor with amenities in which cultural industries (with immobile products) play a central role. The large scale or lumpiness of the amenities makes a strong involvement of the government inevitable. The constraints imposed by the public–private partnership necessitated the inclusion of other elements (office space, hotel accommodation, restaurants and shops, apartments) in high densities to make the project profitable. The city, right from the start, emphasised the need for architectural quality to give the Oosterdokseiland a distinct identity and the master plan was designed by Erick van Egeraat (Van Rooy 2003). The combination of functions has implicitly shifted the focus from creating a *museum cultural district* to creating a *metropolitan cultural district*.

Whereas the Oosterdokseiland was something of a black hole in the city where almost no one went, it has now become part of the itineraries of many inhabitants and visitors of Amsterdam. The beautiful Public Library with its restaurant on top giving a fascinating view of the old city centre of Amsterdam, in particular, has turned out to be crowd puller. As such, the Public Library and its restaurant are the successor of the temporary housing of the Amsterdam Museum of Modern Arts or Stedelijk Museum and its restaurant on top, which were early pioneers on the Oosterdokseiland and turned it already into a cultural anchor able to attract large numbers of visitors. The temporary housing of the Stedelijk Museum that first put the Oosterdokseiland on the mental map of people was not so much planned as part of a lucky coincidence. The famous Stedelijk Museum was looking for temporary space while awaiting the renovation of the original building in the southern part of the city. In 2004, the Museum set up shop on the ground floor of the former postal distribution centre, a large high-modernist building ('one of the highlights of late 1960s functionalism' (Lebesque 2006: 155)), and this became the Stedelijk Museum CS. A turnaround was triggered, especially when on the top floor, a post-modern café and restaurant with breathtaking view of the old city opened. The floors in between, meanwhile, became occupied by small firms in the cultural industries: advertising agencies, web-design firms, architectural practices, photographers, etc. About 70 firms with about 500 workers located in the building attracting a lot of attention (*NRC-Handelsblad* 2008).

From a desolate place, the Oosterdokseiland thus became the place to be for all kinds of creative activities looking for a relatively cheap but attractive high-urban atmosphere. This was, however, temporary as the Post CS building had to be refurbished for its new office function and both the museum and the cultural industries firms had to leave in the summer of 2008. By now, however, the meanwhile completed Amsterdam Public Library 'or Centrale Bibiotheek – with sleek, modern architecture that is stunning against the old waterfront buildings it fronts … [and with] 600 free computer terminals and the seventh-floor restaurant with a panoramic city view' (Greenfield 2009) and the Amsterdam Conservatory (music school) are attracting large numbers of visitors and the Oosterdokseiland is indeed fulfilling its role as a cultural anchor.

The present success of the Oosterdokseiland can be attributed to its central location, the strength of the cultural anchors (Public Library, music hall, music school), and the quality of the architecture. The turnaround from a place to avoid to a place to be and to be seen cannot be properly understood, however, without referring to the pioneering role of the Post CS building and the actors who were involved in this. These actors, the real-estate developer who was open to the temporary use of the building, the Stedelijk Museum who wanted temporary housing for exhibitions, the daring entrepreneur who transformed the eleventh floor of the building into restaurant with breathtaking views of the city life, both outside and inside the restaurant, and the large number of workers who created a buzz in and around the place. Amsterdam did already have the critical mass of such actors who were able to appropriate the hardware in a meaningful way.

Case 2: Accommodating Cultural Industries

Perhaps most significantly, the Eastern Docklands, unlike London's Canary Wharf, is a tourist attraction. I always make time to the pillarbox-red 'Python Bridge' between the Sporenburg and Borneo Islands, ogle the extraordinary apartment blocks, with names such as 'The Whale and Hope', and 'Love and Fortune' and find out what's new on this rapidly evolving city-scape (Bergmans 2007: 119).

The city made plans for a cultural centre on the Oosterdokseiland, the fingers of land in the IJ more to the east were, however, planned as mostly mono-functional residential areas (Lebesque 2006). When the more comprehensive Koolhaas master plan for the whole southern part of the IJ banks fell through, the area was divided into separate plots, each with its own master plan (and architects). What was clear, however, was that density should be very high, that the architectural quality should be high and innovative, but relating to the architectural heritage of the city centre of Amsterdam (Lebesque 2006: 99–101). The KNSM island was initially mainly intended for social housing, especially for new urbanites (two-earner households, mostly without children): 'in terms of programme, the KNSM housing was largely a continuation of policy-driven housing supply, with small, one and two-person units and compact family dwellings with traditional floor plans' (Lebesque 2006: 205). The housing on Sporenburg and on some parts of Java island is mostly private sector and much less supply-driven and more demand oriented (see Table 5.1). Developers were allowed to come up with innovative design and this has resulted in a variegated architectural landscape with surprising designs combining seemingly traditional Amsterdam canal views with apartments with large spacious living rooms.

Table 5.1   Total population and share of private housing in the Amsterdam Eastern Docklands, 2005

Neighbourhood	Total population	Share private housing (%)
KNSM	2,518	46
Java	3,149	34
Sporenburg	2,480	58
Borneo	2,658	41

Source: Amsterdam Regiomonitor (2005).

Both the design of the housing on Java island and on Sporenburg have attracted much attention from the architectural field (De Botton 2006). The bridges connecting Sporenburg, designed by Rotterdam architectural practice West 8, are even famous. According to the *Wallpaper* City Guide* (2006: 69) for Amsterdam: 'One of the keys to the overall success of the development were the three bridges that connect the differences areas and provide a unifying architectural feature. The work of architects West 8, the bridges have become destinations in their own right'. The housing in combination with the spectacular views of low Dutch skies and the IJ River now frequently form the backdrop of television series. This outspoken architectural identity, moreover, has made living there very popular and these neighbourhoods, which were completed around the year 2000 or even later, are very well able to compete with the older, more central parts of the city.

Attracting high-skilled workers nowadays in the digital era also means attracting business. Although the neighbourhoods were mainly intended as residential neighbourhoods, recent data gathered to construct the Amsterdam Regiomonitor show that the fairly recently completed Eastern Docklands are teeming with economic activities (see Table 5.2). In 2008, no less than 1,802 firms were registered in the Eastern Docklands, which means that there are 10.1 firms for every 100 inhabitants in this predominantly residential area. A substantial part of these economic activities match the profile of the digital city as described above: small-scale, home-based businesses, notably in the creative industries (advertising, publishing, writing, the performing arts, the visual arts, architectural and other design), working for non-local markets and probably strongly dependent on ICT for their production. About one-third of the total number of firms can be found in these creative industries. It seems, therefore, that the residential parts of the Eastern Docklands in Amsterdam are actually an incubator for post-industrial cottage industries. As yet, we do not have any in-depth information on how these firms function and how caring, work and leisure are combined in time and space. We also do not know to what extent these small firms are part of local networks. If that is the case, the Eastern Docklands would be moving in the direction of an industrial cultural district. It appears, however, that here we indeed see the emergence of new patterns of economic activities that are characteristic for digital cities. Our research project on economic activities in residential neighbourhoods, which will be completed in 2012, should shed more light on this.

Table 5.2   Economic activities in the Amsterdam Eastern Docklands; total number of workers, firms and share of creative industries, 2005

Neighbourhood	Total number of workers	Total number of firms	Number of creative workers	Number of creative firms	Share of creative workers (%)	Share of creative firms (%)
KNSM	18	259	231	91	45	35
Java	58	238	142	83	31	35
Sporenburg	96	151	78	51	26	34
Borneo	97	162	65	50	22	31

Source: Amsterdam Regiomonitor (2005).

Lessons from Amsterdam

> Within a relatively short period of time, the South bank has been transformed into a mixed zone of housing, commercial activities and recreation. (*Wallpaper* City Guide* 2006: 6)

The departure of large-scale industrial and logistical activities left gaping holes in many cities. These spaces are now being re-integrated in the city. The economic profile of the city, however, has fundamentally changed due to digital technology and globalisation. New economic activities, new uses of space, new ways of organising work, new spatial articulations are emerging in cities making them distinctly different in some key respects from their industrial predecessors. To be successful, urban regeneration projects have to match the larger transformation processes of the city by creating opportunities for the new activities and the new uses of space. Cultural industries can be used as instruments for regeneration, but there are clear caveats. One cannot create an industrial cultural district out of the blue, though one can help to create conditions as shown by the Eastern Docklands. One can use large cultural anchors to transform a particular area as has been proved in Bilbao, London, Paris and now also on the Oosterdokseiland in Amsterdam. These recent transformations of the Amsterdam Eastern Docklands into a buoyant part of the old city might have some larger lessons to offer.

Many of these lessons were already formulated by Jane Jacobs (1961) when she criticised the modernist re-ordering of New York. In the digital city with its small firms, its embeddedness in the locality, its importance of public and semi-public (third) spaces, its dependence on face-to-face contacts and its deep integration of work, caring and leisure, these lessons are arguably more poignant then ever. I have summarised these lessons under five headings.

Plan for Flexibility: Leave Spaces Open

It now seems that the Eastern Docklands are an example of a successful transition from an industrial to a post-industrial use. The carving up of the area into several plots with their own master plans created diversity. High densities helped to create an urban atmosphere. Notwithstanding these evident qualities, the Eastern Docklands are also victims of planners' hubris. The plans were notably designed to provide housing for couples without children (what were once called yuppies). Human nature does not, however, always follow the patterns envisaged by planners. Many of the couples did eventually have children and now the neighbourhoods suffer from a lack of childcare facilities and schools. Plans for transforming large areas inevitably suffer from a lumpiness determined by infrastructural, organisational and financial demands. One cannot, for example, construct a tramway or a metro in a piecemeal way. Within these constraints try to keep the plans as open as possible on the level of the individual housing (which seem to achieved in parts of the Eastern Docklands), but also on the level of the street and the neighbourhood if more and other functions should be needed. If indeed an industrial cultural district is developing in the Eastern Docklands, spaces for face-to-face contacts and many office buildings are needed to accommodate the local businesses.

An open, oblique and adaptive approach to planning as promoted by people from Charles Lindblom (1959) to Jane Jacobs (1961) and from James Scott (1998) to John Kay (2010) is still needed, perhaps even more so now as ways of production, consumption, lifestyles, uses of space and time schedules become more diverse. No planner will ever have the information necessary for top-down planning and a modest attitude, leaving as many spaces at different scale levels open to be filled in (literally and symbolically) by the users themselves as possible, might avoid planning disasters and create places and neighbourhoods which will last for ages, just like the historical city centre of Amsterdam itself.

Plan for Quality: Architectonic Quality will Pay for Itself

Both the Oosterdokseiland and the Eastern Docklands have underlined the importance of high-quality architectural design. It helps to give a new neighbourhood a distinct identity. The architectural quality has to be realised on several levels: from the master plan to the interior of the individual houses. Large structures as bridges and public libraries can function as signature buildings. The droves of architectural tourists who go to the Eastern Docklands and its popularity as a backdrop for commercials and television series, and, moreover, its popularity on the Amsterdam housing market are testimony to the overall architectural quality of this area.

Co-operate with Private Actors, but Preserve Openness

These large regeneration projects can usually only be realised in close collaboration with the private sector. One should, however, be watchful and defend the public interest. In the Eastern Docklands, one essential element of urbanity is almost lacking: a fine-grained mix of housing and retail. The real-estate developer in conjunction with a large retail firm was able to draw up a zoning plan which does not allow dispersed small-scale retail and concentrates nearly all (large-scale) retail in one specific place. The neighbourhood might be now saved to some extent by the boats on the ubiquitous quays housing cafés and restaurants, however the fact remains that the retail landscape is more like a desert. Public bodies involved in planning should not allow private actors to dictate the zoning plans.

Stimulate Self-organisation and the Importance of Local Actors

The early success of the Oosterdokseiland was clearly partly driven by the lucky presence of dynamic entrepreneurs who were able to see, on the one hand, which opportunities were on offer in this area and, on the other, what needs existed among the diffuse group of up and coming creative workers. The developer made space for them, literally, and a cultural anchor appeared on the radar. As with space in the literal sense, regeneration plans should leave institutional space open for new actors who as gatekeepers are much more in touch with new economic activities.

No Panacea for Every City: Dependent on Urbane Population

The transformation of the Amsterdam Docklands seems to chime with the shifts in the socio-economic base of the city. Amsterdam is the city that now attracts the largest number of highly educated people in the Netherlands. Graduates from university towns as Leiden, Utrecht, Rotterdam, Maastricht and Enschede move to Amsterdam in droves in search for jobs but also looking for a cosmopolitan atmosphere and a wide range of amenities (Tomesen 2011). These new urbanites are attracted by what's on offer in the Amsterdam Eastern Docklands: an urban environment that enables them to combine working, leisure and care in new ways. They are able to go to a concert, a restaurant, a café, a museum, a host of other third spaces and there is also a fertile environment not only to find a job but also to start a business and tap into the rich networks of the nearby city centre. Without them, the Oosterdokseiland would never be successful. Amsterdam has the critical mass of entrepreneurs, the urbane people with financial *and* cultural capital, and an extensive local production system in the cultural industries. All plans for new urbanity in old cities can only work if there is a sound basis of old urbanity.

References

Bakker, J.D. 2004. *Harmonie aan de IJ-over. Onderzoek naar de clustering van culturele instellingen op het centrale deel van de Zuidelijke IJ-oever.* Doctoraalscriptie. Amsterdam: Universiteit van Amsterdam.

Beauregard, R.A. 2002. New Urbanism: Ambiguous Certainties. *Journal of Architectural and Planning Research* 19(3): 182–94.

Bergmans, J. 2007. Laid Back, Switched On, Cutting Edge. *Financial Times/Life and Arts*, 29–30 September, 11.

Blokker, B. 1995. De desolate stadsrand. *NRC-Handelsblad*, 22 July.

Butler, T. 2007. Re-urbanizing London Docklands: Gentrification, Suburbanization or New Urbanism. *International Journal of Urban and Regional Research* 31(4): 759–81.

Caves R.E. 2000. *Creative Industries: Contracts between Art and Commerce.* Cambridge, MA: Harvard University Press.

Cheshire, P.C. 2006. Resurgent Cities, Urban Myths and Policy Hubris: What We Need to Know. *Urban Studies* 43(8): 1231–46.

Clark, P. 2009. *European Cities and Towns 400–2000.* Oxford: Oxford University Press

Currid, E. 2007. *The Warhol Economy: How Fashion, Art and Music Drive New York City.* Princeton: Princeton University Press.

Currid, E. and Connolly, J. 2008. The Geography of Advanced Services and the Case of Art and Culture. *Annals of the Association of American Geographers* 98(2): 414–34.

De Botton, A. 2006. *The Architecture of Happiness.* London: Hamish Hamilton.

Florida, R. 2002. *The Rise of the Creative Class: And How it's Transforming Work, Leisure, Community and Everyday Life.* New York: Basic Books.

Friedman, T. 2005. *The World is Flat. A Brief History of the Twenty-First Century.* New York: Farrar, Straus and Giroux.

Gomez, M.V. 1998. Reflective Images: The Case of Urban Regeneration in Glasgow and Bilbao. *International Journal of Urban and Regional Research* 22(1): 106–21.

Greenfield, B. 2009. The Amsterdam of Playgrounds and Pancakes. *New York Times*, 20 September.

Hall, P. 1998. *Cities in Civilization*. New York: Pantheon.

Hall, P. 1999. The Future of Cities. *Computer, Environments and Urban Systems* 23: 173–85.

Hall, P. 2000. Creative Cities and Economic Development. *Urban Studies* 37(4): 639–49.

Hall, P. and Pain, K. 2006. *The Polycentric Metropolis: Learning From Mega-City Regions in Europe*. London: Earthscan.

Hobsbawm, E. 1996. *The Age of Extremes. A History of the World, 1914–1991*. New York: Vintage Books.

Hohenberg, P. and Hollen Lees, L. 1995. *The Making of Urban Europe 1000–1994*. Cambridge, MA: Harvard University Press.

Hutton, T.A. 2004. The New Economy of the Inner City. *Cities* 21(2): 89–108.

Jacobs, J. 1961. *The Death and Life of Great American Cities*. New York: Random House.

Judt, T. 2005. *Postwar: A History of Europe since 1945*. London: William Heinemann Ltd.

Kay, J. 2010. *Obliquity. Why Our Goals Are Best Achieved Indirectly*. London: Profile Books.

Kloosterman, R.C. 2004. Recent Employment Trends in the Cultural Industries in Amsterdam, Rotterdam, The Hague and Utrecht; A First Exploration. *Tijdschrift voor Economische en Sociale Geografie* 95(2): 243–62.

Kloosterman, R.C. 2008. Walls and Bridges: Knowledge Spillover between 'Superdutch' Architectural Practices. *Journal of Economic Geography* 8(4): 545–63.

Kloosterman, R.C. and Lambregts, B. 2007. Between Accumulation and Concentration of Capital: Comparing the Long-term Trajectories of the Dutch Randstad and London Urban Systems. *Urban Geography* 28(1): 54–73.

Kurpershoek, E. 2007. Scheepsbouw, havenregio Amsterdam en herontwikkeling van de IJ-oevers, in *Geschiedenis van Amsterdam: Tweestrijd om de hoofdstad, 1900–2000*, edited by P. de Rooy. Amsterdam: Veen, L.J., 356–97.

Lebesque, S. 2006. *Along Amsterdam's Waterfront: Exploring the Architecture of Amsterdam's South IJ Bank*. Amsterdam: Amsterdam Development Corporation.

Le Goff, J. 1997. *Pour l'amour des villes: Entretiens avec Jean Lebrun*. Paris: Textuel.

Ley, D. 1996. *The New Middle Class and the Remaking of the Central City*. Oxford: Oxford University Press.

Lindblom, C.E. 1959. The Science of 'Muddling Through'. *Public Administration Review* 19(2): 79–88.

Mak, G. 1992. *De Engel van Amsterdam*. Amsterdam: Atlas.

McNeill, J.R. and McNeill, W.H. 2003. *The Human Web. A Bird's-Eye View of World History*. Chicago: University of Chicago Press.

NRC-Handelsblad. 2008. Stedelijk Museum gaat zwerven door de stad, 12 August.

Pratt, A.C. 1997. The Cultural Industries Production System: A Case Study of Employment Change in Britain, 1984–91. *Environment and Planning A* 29: 1953–74.

Raban, J. 1974. *Soft City*. London: The Harvill Press.

Rantisi, N.M. 2004. The Ascendancy of New York Fashion. *International Journal of Urban and Regional Research* 28(1): 86–106.

Rykwert, J. 2002. *The Seduction of Place: The History and Future of the City*. New York: Vintage Books.

Santagata, W. 2002. Cultural Districts and Property Rights and Sustainable Economic Growth. *International Journal of Urban and Regional Research* 26(1): 9–23.

Sassen, S. 1991. *The Global City. New York, London, Tokyo*. Princeton: Princeton University Press.

Scott, A.J. 2000. *The Cultural Economy of Cities: Essays on the Geography of Image-Producing Industries*. London, Thousand Oaks and New Delhi: Sage Publications.

Scott, A.J. 2004. Cultural-Products Industries and Urban Economic Development: Prospects for Growth and Market Contestation in a Global Context. *Urban Affairs Review* 39(4): 461–90.

Scott, A.J. 2005 *On Hollywood: The Place, The Industry*. Princeton and Oxford: Princeton University Press.

Scott, A.J. 2006. Creative Cities: Conceptual Issues and Policy Questions. *Journal of Urban Affairs* 28(1): 1–17.

Scott, A.J. 2007. Capitalism and Urbanization in New Key? The Cognitive-Cultural Dimension. *Social Forces* 85(4): 1465–82.

Scott, A.J. 2008. *Social Economy of the Metropolis: Cognitive-Cultural Capitalism and the Global Resurgence of Cities*. Oxford: Oxford University Press.

Scott, J.C. 1998. *Seeing like a State: How Certain Schemes to Improve the Human Condition have Failed*. New Haven: Yale University Press.

The Economist (2008). Nomads at Last. 10 April.

Tomesen, R. 2011. Het platteland wordt echt platteland, de stad echt stad. Nederland is er bij gebaat dat Amsterdam groter is. Aantrekkelijk Amsterdam. *De Pers*, 14 March.

Trip, J.J. 2007. Assessing Quality of Place: A Comparative Analysis of Amsterdam and Rotterdam. *Journal of Urban Affairs* 29(5): 501–17.

Van Rooy, M. 2003. De palazzi van het Oosterdokseiland. *NRC-Handelsblad*, 5 November.

Wallpaper* City Guide. 2006. Amsterdam. London: Phaidon.

Zukin, S. 1995. *The Cultures of Cities*. Oxford: Blackwell Publishers.

Zukin, S. and Maguire J.S. 2004. Consumers and Consumption. *Annual Review of Sociology* 30: 173–97.

Chapter 6

From the Old Downtown to the New Downtown: The Case of the South Boston Waterfront

Susanne Heeg

Introduction

According to Ilse Helbrecht and Peter Dirksmeier (Chapter 1, this volume), we are now seeing a new form of centrality in a globalised world. New downtowns, new urban spaces within the established framework of older cities, are appearing in order to meet the demand for centrally located urban life. The phenomenon of the new downtown is, according to these authors, a result of the renewal of inner cities, as a result of which it has become possible to integrate convertible spaces near the city centre back into the territory of the city. The high 'demand for the city centre' thus motivated the repurposing of industrial areas, which has made it possible to create new urban spaces which augment the old city centre. As a result, the new downtown is not just the result of the fall of the old city centre, but rather is based on the renewed valuation of the old downtown. The new downtown does not arise in competition with the old inner-city territory, but resembles more the uncovering of a new possibility arising out of the increased value attributed to the old downtown. The increased valuation of *both* areas can then be attributed to the rediscovery of the quality of life available in urban centres on the part of residents, city planners and the real-estate industry.

The following will make a case study of the South Boston Waterfront in order to analyse the conflicting interests which come into play when the creation of a new downtown is attempted. The South Boston Waterfront is a large-scale conversion project, directly bordering on the Boston City CBD.[1] This project can serve as an example on the basis of which to consider whether the revaluation of the old and

---

1   The Central Business District is treated here as a part of the old downtown. Within the CBD there is a concentration of office towers with the corresponding jobs and service infrastructure. The usage structures connected to it are those which are primarily concentrated around working hours. A real downtown is normally more complex, including streets of shops and residential areas which offer restaurants and recreation. It is these parts of the city which are perceived as lively and diverse, since the use of the space can potentially cover all 24 hours of every day.

the new downtown augment each other or, instead, result in a clash between the two areas. This will be examined in the context of the privileging of the interests of private industry over those of the residents, which is typical of but not limited to American cities. Specifically, the challenges which go along with the development of a new downtown in situations where the city planners are tasked with providing the conditions for the development of the area, but not with intervening in the property-owners' wishes for the use of their property, will be analysed. So far, the development of the new downtown in Boston has proceeded with generous public start-up funding in order to encourage private property investment. Contrary to the usual accounts offered by the media, the development of new downtowns does not take care of itself, but requires intensive publicly funded interventions and investment in order to warm up, attract interest and get started. Only when the development and management of these new locations can be made attractive to potential users and promise high profits for property investors is the interest of the private real-estate industry in as-yet unconventional locations sparked. In order to achieve this, the city government must make advance investments in providing infrastructure, the recycling of old properties, direct and indirect subventions, etc. However, if, once private developers' interest has been sparked, city governments do not exercise regulatory authority, there is a danger that property owners will pursue their own individual interests, which generally involve the realisation of office projects in order to turn a high profit. This is particularly the case with areas near the city centre and bordering directly on the CBD. Expanding office property appears to be an advantageous usage given the connection between the locations. In sum, the pursuit of individual interests will not lead to a lively downtown which attains the goals formulated in public debates (new spaces for recreation and social contacts) but instead will lead to an expansion of the CBD. As a result, the new downtown could easily become a mere satellite of the CBD, full of people during working hours, but abandoned in the evenings. The proximity of the old downtown is an opportunity for the development of a new downtown, but also a danger, if too dense a concentration of office developments is allowed. An urban planning scheme which is oriented to principles of laissez-faire and enabling of investment appears, as Boston demonstrates, to make sense only on the condition that few or no real-estate interests exist in a given area. If, however, an area becomes the focus of real-estate interests, it is necessary to take regulatory measures to prevent total urban mis-developments resulting from the sum of individual economic decisions.

Successful public financing of revaluation schemes can thus be a double-edged sword, as I have argued on the basis of the South Boston Waterfront. One must ask whether a new downtown can develop at all out of extensive public investment and city planning which intervenes selectively at best in private real-estate trade. Developments up to this point suggest that the new downtown is bringing with it an expansion of the Central Business District, as part of the old downtown, into the South Boston Waterfront. This raises several questions with regard to the discussion initiated by Helbrecht and Dirksmeier on the topic of new downtowns. Can the new downtown actually be distinguished from the old downtown?

Are there any efforts towards producing a new urban space with a high quality of life? How do those involved (city planners, residents, real-estate investors) interpret the emphasis on urban qualities in public discussions? Who has the right to determine which plans will be realised in which places? This chapter argues that, despite a high valuation of urban spaces which include a large proportion of residential and recreational facilities, other sorts of spaces are realised in the vicinity of city centres. These 'other' spaces result from efforts to make real estate profitable, which in turn leads to the expansion of office space into the new downtown. Thus it is necessary, in the development of new downtowns, to consider the significance, interests and role of the real-estate industry as a formative agent within the city.

In the following section I will summarise the development of downtown Boston with particular attention given to those areas near the water. This is intended to show that there were two historical precedents to the development of the South Boston Waterfront, consisting of a phase of total demolition as well as one of renewal with a view to architectural and touristic upgrading of inner-city areas. In both phases, as well as in the present phase, there is no significant difference in the goal towards which public monies are being directed with regard to the development of this area: it is a matter of preparing locations for private real-estate investment. A difference can however be seen in the degree of interest obtaining on the part of the real-estate industry. The public revaluation measures in the first two phases are reactions to large-scale processes of socio-economic changes in the post-war era and represent attempts to establish new patterns of usage. In particular, the foundation was laid for the revaluation of the entire city centre with its orientation to tourism, festivalisation and mallisation. This in turn was the precondition for city planners and real-estate investors turning their attention to the South Boston Waterfront. Their hope was that a new downtown could be developed close to the CBD. The ambitions of real-estate investors and those of city planners however are in tension with each other. Will downtowns result which compete with each other or will they augment each other? The second and third sections below analyse the context of these problems, and efforts at revaluation and their results, in the old downtown from a historical perspective. My main focus is on the role of urban planners as facilitators, but not regulators, of private profit-oriented efforts.[2] The final chapter addresses the question of whether a new downtown is in fact developing in this case.

---

2   This section on the South Boston Waterfront is based on a large number of interviews with those involved in the development of the South Boston Waterfront which were conducted between May and September 2005. This group includes city planners, funding bodies, consulting firms, investors, project developers, neighbourhood organisations, residents and non-governmental organisations. This section is also based on an analysis of reporting on this topic in the American press (especially *The Boston Globe*, *The Boston Herald* and *The Boston Business Journal*) as well as non-commercial newsletters and pamphlets (so-called gray literature).

The Post-war City Centre: Suburbanisation and Demolition

Like many American cities in the post-war era, Boston was the scene of a rapid
decline which manifested itself in 'white flight', the dispersal of services and
industry, and a lack of investment in existing buildings. These changes affected large
portions of the city but were mainly discussed as they applied to the city centre and
adjoining areas. As a result, efforts were soon undertaken to revitalize downtown
Boston and the waterfront. Boston shares these features with the city of Baltimore
and both cities were recognised as American poster children for the revitalisation of
the city centre and revaluation of the waterfront (Frieden and Sagalyn 1990; Ward
2002), although Boston gained more recognition for revitalisation and Baltimore
more for the development of the waterfront. Without exception, all efforts at
revaluation are a reaction to suburbanisation and deindustrialisation and thus to
falling tax revenues. However, different phases of revaluation can be identified. In
Boston, the main milestones in the process of revaluation are the demolition and
total overhaul of the West End (started in 1958) on what is today the site of the
Government Center (opened in 1967) as well as the phase of attracting tourism
by means, for example, of the Faneuil Market Hall (opened in 1976) and Rowe's
Wharf (completed in 1987). The individual reconstruction projects are adjacent to
each other and represent an expansion and upgrading of the downtown area. This
is also the case with the South Boston Waterfront, which is the last remaining area
available for expansion adjacent to the CBD.

Until revaluation measures for the South Boston Waterfront were first
discussed, it was neither a part of the CBD nor a part of the old downtown. This
is connected to the fact that the South Boston Waterfront was characterised by
usage frameworks for industry and shipping up to the 1980s. In addition, until the
decision was made to tunnel under Interstate 93, which was a elevated highway
on the edge of the CBD, the interstate formed a barrier between the CBD and the
South Boston Waterfront.

Boston has experienced suburbanisation since the 19th century. Suburbanisation
was strictly limited, however, by the space available between Boston and
neighbouring industrial cities. When the Boston suburbs expanded, they thus filled
in the space between the city and industrial sites in the backwaters (Bluestone and
Stevenson 2000: 77). Suburbansation only became a problem for Boston, as it did
for other American cities, in the years after the Second World War. Since then,
suburbanisation has also brought with it the dispersal of business and of wealthy
segments of the population and was thus the precursor to increasingly acute urban
problems resulting from the loss of tax revenue and the absence of investment in
parts of the existing buildings and residential areas of the inner city. The creation
of attractive urban spaces was intended to stop this development. A decrease in
white flight and the dispersal of businesses can only be observed starting in the
1980s (see Table 6.1).

A decisive factor which contributed to suburbanisation in the United States was
the increase in prosperity. This allowed a larger number of private households to

own their own car and to live in their own home at some distance from their place of work. The significance of the expansion of transportation infrastructure for this process should not be neglected. State freeway programmes contributed to the establishment of a network of highways between major US cities. These cities thus became more easily accessible from more distant locations (Frieden and Sagalyn 1990: 25–6). An additional factor which contributed to suburbanisation was the public bonds provided by the Federal Housing Administration for the construction of family homes. This meant a risk-free mortgage for the banks, a new market for the real-estate industry, better conditions for borrowers and so-called white flight for the cities (Bluestone and Stevenson 2000: 81; Mollenkopf 1983: 41). It was primarily white Americans who were in possession of adequate income to own a house in the suburbs and to improve their material standing. They tried by this means to escape from the worsening conditions in the inner cities, but their move exacerbated the erosion of those very conditions.

Table 6.1   Population development in Boston and the neighbourhood of South Boston

Census Year	Boston and the Islands (abs.)	South Boston (abs.)
1950	801,444	55,670
1960	698,081	45,766
1970	641,071	38,488
1980	562,994	26,666
1990	574,282	29,467
2000	589,141	29,959

Source: Boston Public Library (2006).

Like other cities, Boston was affected by white flight and the dispersal of services and industry. Industrial dispersal was further encouraged by the combination of opportunities to expand spatially and the low cost of property on the periphery. According to Mollenkopf (1983: 23), the suburbanisation of industrial jobs was more pronounced than that of the populace. However, not only shifts, but also newly established industries took place more frequently on the periphery than in the city along with the migration of an economically active populace. Finally, services followed their customers out to the periphery. Examples of this latter trend are retail, medical services and customer-oriented services of banks, insurance companies, etc.

Boston was thus characterised by a shrinking population, deindustrialisation and, with it, a loss of industrial jobs in the post-war era. At the same time, private investments and maintenance sank in large parts of the city centre, so that residential buildings and other buildings fell into disrepair. Not one office building was built in the city centre between 1920 and 1959. 'Boston's total real estate valuation declined 28 percent from a pre-Depression high of $1.8 billion to $1.3 billion in 1960' (Mollenkopf 1983: 145–6).

These socio-economic changes could be seen in the state of the buildings in the CBD. When the economic mix in the CBD[3] proved to be no longer feasible, business-related buildings such as warehouses, stores and older office building lost value. Similar tendencies towards devaluation were also observed in residential areas around the circumference of the CBD. Examples of the latter are the West End, Charlestown, South End and South Boston.[4] Approximately half of the population lost to Boston between 1950 and 1970 came from inner-city neighbourhoods which became no-go areas for large sections of the population. When the white population migrated, many minority households were established in their place, which encouraged white flight and anxiety about the loss of property value (Kennedy 1992).

Buying inner-city properties was considered a poor investment into the 1970s. Typical real-estate projects were shopping malls and residential developments in the suburbs. For Boston, as for other cities, these developments meant a loss of economic foundation and tax revenue. Unlike in Germany, property tax on residential and business properties represent a central source of municipal income.[5] Since the tax revenue is calculated on the basis of the value of the buildings, the absence of investment in the built environment meant the devaluation of inner-city locations and therefore a decrease in tax revenue.

At several points in the post-war decades, the city has faced outright fiscal crisis. In 1959, the city had the lowest bond rating of any city with a population of over 500,000; this nadir was the culmination of years of economic decline and disinvestment in the built environment (Bluestone and Stevenson 2000: 102).

At this point there was a threat of a vicious cycle of migration and inner-city decay, which would motivate further migration and lead to further loss of tax revenue. Necessary improvements to the infrastructure were made to wait and municipal social services were reduced. This provides an explanation for

---

3   The terms CBD and downtown are often used synonymously. In this text they are differentiated as noted in footnote 1 above. In Boston, the term CBD refers to the area around the Government Center (hence central) and the Financial District. Downtown includes the CBD and adjacent neighbourhoods which have gained in significance as a result of renovation. These include Back Bay, West End, North End, Beacon Hill and the northern section of the South End (especially the Theatre District and Chinatown).

4   'The 1950s provided a picture of commercial and industrial decline with the increasing presence of "blighting" lower-class minority groups and a lack of business confidence and investment even in the central business districts' (Mollenkopf 1983: 144). In contrast, the suburbs experienced above-average economic growth.

5   Property tax revenue in Boston in 1999 was 50.3% of municipal income. In 2006 it had risen to 58.9% (BMRB 2006: 2). In a study of the correlation between economic development and property tax in American cities it was noted that almost all local governments examined had used land as a revenue-generating device (Chapman et al. 2005).

the reasons why the implementation of inner-city reconstruction projects took on high priority. However, these projects were only incidentally intended to stabilise or preserve the downtown area as a place for interaction, communication and encounter. More centrally, the measures implemented were intended to stabilise the economic function and residential situation of the area. In fact, the efforts at revaluation destroyed downtown Boston as a place of communication of less prosperous and minority households.

Counter-measures

In many American cities, urban alliances formed in the post-war era in reaction to a problematic situation in the inner cities in order to formulate an agenda for local development and to concentrate efforts on necessary changes. The business community in Boston, which consisted of a close network of important companies, large banks and established legal practices, constructed various facilities[6] in this period and organised events (such as the College Citizen Seminars in Boston College) aimed at formulating the Central City Agenda (Mollenkopf 1983: 154). In the course of discussions between municipal and business representatives over 10 years, consensus was reached that inner-city decay should be confronted by means of reconstruction. In the General Plan for Boston of 1950, it was agreed that 27 hectares of the total 388 hectares of the downtown area would be reconstructed. This was synonymous with demolition and later rebuilding (Keyes 1969: 26–7; Mollenkopf 1983: 144). The Boston Redevelopment Authority (BRA) was founded in 1957 to carry out this task and was to combine the necessary economic, real estate and planning know-how. The BRA, which was described in the literature as an exceptionally powerful, if not all-powerful, city planning office (Horan and Jonas 1998; Mollenkopf 1983: 139), was responsible for the financial and practical implementation of individual revaluation projects.

The first project which was completed and which attained national recognition was the renovation of the West End, a neighbourhood to the north-west of the CBD, at the end of the 1950s (McQuade 1966). Renovation was equated with tearing the majority of existing buildings down, a process which began in 1958 and ended in 1960.[7] Demolition was justified with reference to the large proportion of 'cheap housing' (Jakle and Wilson 1992: 312). So-called cheap housing was considered an indicator for a low building standard and a bad hygiene situation (Keyes 1969: 27).

---

6   The most important organisation in this regard is *the Vault* – named after the vault of the *Boston Safe Deposit & Trust Co.*, where a circle of politicians and businessmen met in order to make agreements about the means and ends of the urban renewal process.

7   'The neighborhood's 7,000 residents were notified in April 1958 that the city was taking their dwellings under the power of eminent domain. The bulldozers began rolling in June 1958: by November, 1,200 of the West End's 2,700 households were gone. By 1960, only five of its buildings were left standing, including Massachusetts General Hospital and the old Charles Street Jail' (Walker 1999).

John Mollenkopf (1983: 157) quotes Mayor Hynes, who supported the project: 'The only way that decay and blight may be uprooted … is by a complete physical change in the affected neighborhood or area'. The next major project was in the red-light district, south-east of the West End. The Government Center, Boston's city hall, was built on that location (see Figure 6.1) using public money and intended as a signal for the revaluation and changed usage of the area.

These large-scale renovation projects were made possible by the Federal Housing Act of 1949, better known as the Urban Renewal Act. The goal of the programme was the removal of urban blight, that is, visible signs of decay, poverty and the deterioration of buildings in the inner city. To motivate building projects, Chapter 121A provided for tax relief and changes in the use of a property, which allowed for different types of buildings and building use. In order to qualify, the location had to have been identified as suffering from 'urban blight'.

Figure 6.1   View of the Government Center with the harbour in the background
Source: Wikipedia (2009a).

The definition of blight was based on social and architectural factors and was highly controversial, because the cohesion of neighbourhoods was ignored in favour of removing visible signs of poverty (Weber 2002: 179). Frieden and Sagalyn refer to this procedure as 'sanitising the city': 'Soon city renewal directors were searching for "the blight that's right" – places just bad enough to clear but good enough to attract developers' (Frieden and Sagalyn 1990: 23). There is broad consensus in the geographical, political and sociological literature that the explicit goal of efforts at reconstruction in the post-war period was the economic revaluation of the city centre (Frieden and Sagalyn 1990; Kennedy 1992; Mollenkopf 1983; O'Connor 1993). The projected new city centre was to move with the times and replace an outdated mix of retail and small businesses, that is, to become available for new functions such as administration, retail and business-oriented services. Access was to be improved by the expansion of the transport infrastructure. At the same time, identifying a need for renovation in the area encircling the CBD (the West End, South End, Charlestown, etc.) aimed at social upgrading in favour of the middle class, for whom housing was to be built. New buildings seemed necessary in order to meet these requirements. Public investments were a means to this end. In the case of the West End, public monies were used to raze old buildings and thereby destroy neighbourhoods in order to rebuild for other residents on what is now the site of the Government Center.

It was primarily areas with a large proportion of minority residents which were chosen for renovation (Frieden and Sagalyn 1990: 28). The selective inclusion of neighbourhoods in such measures confirmed the impression that this was not only about resolving the problems existing with inner-city buildings and supporting the old downtown, but rather it was also about the creation of locations for more prosperous classes and economic activities in order to achieve a larger tax base. This is in line with Thomas O'Connor's argument:

> It would be difficult for the administration to boast of a proud New Boston [during the 1960s, S.H.] when the refurbished downtown business district and Government Center were still hemmed in by a series of run-down old neighborhoods populated by blue-collar workers, low-income people of color, and a depressing collection of the homeless and the disadvantaged. The city's new program was designed to transform those valuable locations into clusters of attractive communities with the kind of shiny new town houses and modern apartments that would bring middle-class families and well-to-do professionals back to the city, where they would make Boston once again an appealing place to live and a convenient place to work. (O'Connor 1993: 214–15)

These rebuilding projects were made possible by a large federal programme for urban modernisation and renewal. Federal funding allowed municipal rebuilding organisations to buy up 573 hectares of inner-city property between 1954 and 1969. In that period, 1,600 new projects were completed.

Federal aid for cities rose from 0.3 billion dollars in 1957 to a total of 9.3 billion dollars at its highpoint in 1977. Boston received about 300 million dollars from 1957 to 1972, more than any other American city. In Boston, Ed Logue, who was the director of the Boston Redevelopment Agendy[8] at the time, was able to acquire one-quarter of the city's total surface area, which was home to half of the city's population, by means of federally funded renewal programmes (Mollenkopf 1983: 165). These funds also made it possible to involve the residents and to engage in a long and costly planning process. This became necessary when the rebuilding of the West End encountered strong criticism. In general, city planning was understood to be a proactive task in this phase. In the face of the perceived decline, government was expected to intervene in order to make certain areas attractive to private business again. Government intervention and investment served to restore the areas in question sufficiently so that property owners recognised the value of their real estate and made new investments.

Renewal of the City Centre in the Entrepreneurial City

An assessment of the revaluation and rebuilding policy of the 1960s requires distinguishing between economic, architectural and social factors. It can certainly be conceded, however, that the policy successfully met its goals (economic improvement, stabilisation of real-estate prices, and making inner-city housing more attractive to middle- and upper-class households), over the course of 20 to 30 years and put downtown back on the map as a place to live and work. This success, which manifested itself in a building boom and economic expansion, especially in the financial and business services sectors, only became apparent at the beginning of the 1980s. In the 1960s and 1970s, the area first went through a valley of tears, so to speak. In that valley of tears, there were developments which spoke against pursuing the policy further in the form of the rebuilding of large areas. These were:

1. the increasing difficulty of carrying out rebuilding projects in the face of organised resistance by community groups;
2. the lack of demand for inner-city property;
3. the cessation of federal funding as a result of the opposition of the Republican Party to the urban renewal programmes;
4. an economic recession.

---

8   The Boston Redevelopment Authority (BRA) is the quasi-public agency for economic development and urban planning in Boston with far reaching responsibilities.

This combination of causes led to a change in policy and there was a greater involvement of community groups in decision-making (Kennedy 1992: 198) as well as a transition to businesslike and entrepreneurial approaches which also occurred in other cities (Harvey 1989; Pagano and Bowman 1995: 77; Sagalyn 1989). In Boston this meant more orientation towards the development of the downtown area and new forms of co-operation between the municipality and real-estate interests.

Co-operation with investors was pursued on the basis of centralised government renewal programmes as late as the 1970s. In this phase of total overhaul, areas were prepared for building without previous consultation with potential developers. Often, no interested developers could then be found. It became necessary to develop new strategies when funding from federal renewal programmes could no longer be counted on and problems had arisen with valuing inner-city property. In this new phase, which Frieden and Sagalyn (1990: 133) also refer to as the deal-making phase, municipal representatives began consulting with developers before an area was cleared in order to establish interest. In such consultations, the relevant opportunities, challenges and limitations for all parties became clearer and could be factored into the plans. As a result, the role of the planning offices changed and they became developers and co-investors themselves. Negotiations often resulted in their being allocated certain parts of project development and assessing the possibilities for financial relief or incentives. From that time on, the role of city planning was to facilitate investment on the part of private businesses. This became a realistic strategy once representatives of the real-estate industry had become more interested in investing in the downtown area.

A good example for this is the development of Faneuil Hall Marketplace (Figure 6.2). Faneuil Hall and Quincy Market were decaying buildings adjacent to the Government Center and which, no longer fulfilling their intended purpose, served as a public assembly place and a large market. In contrast to the Government Center, the Faneuil Hall Marketplace project represents the attempt to preserve historic buildings and, by means of festivalisation, to draw renewed attention downtown. To this end, private real-estate investors, especially developers, were included, but only after the municipality had also made investments and done some preparatory work.

Figure 6.2   Faneuil Hall Marketplace and Quincy Market
Source: Wikipedia (2009b).

While initial efforts in the 1960s, under the auspices of Mayor Collins and BRA Director Logue were not particularly successful, Mayor White[9] was able to get development moving again. The sluggishness of development can largely be credited to the fact that, as already mentioned, the inner city was not considered a promising location for retail and restaurant property. The city made some concessions after a long search for suitable projects and ideas in order to stimulate development. These concessions served to reduce the risk to developers and to shore up investments. Municipally owned land was leased to the investors with an initial payment cap. For the developers, leasing was an advantage because it did not require taking on an additional loan in order to purchase land. Reducing the costs of development was necessary, because banks were generally very reserved about financing inner-city projects (Sagalyn 1989).

An innovative scheme for repurposing historical buildings towards recreation, restaurants and gourmet shopping was agreed with the developer. The Faneuil Hall Marketplace development and usage scheme proved to be a success when it was opened in 1976 and it was copied in many other cities. Since then, the preservation and revaluation of historical buildings in the inner city is achieved by means of transforming the buildings into malls focussed on recreation and specialty foods (called Rousification in the United States, in reference to the developer

---

9   Kevin H. White was mayor from 1967 to 1983.

James W. Rouse). Rouse's scheme turned out to be an urban planning hit which contributed to the restoration of downtown areas by means of historicising and staging high-quality and well-controlled urban shopping and recreational spaces. In the wake of this project, apartments were being built again, mainly concentrated in the area adjacent to the waterfront, thus contributing to activity in the downtown area outside of working hours.

This example shows that deal-making entailed offers to developers and investors intended to convince them to invest and to persuade them of the value of certain ways of orienting and implementing rebuilding. The danger in all of this was that there was room to renegotiate in case of changed conditions, such as unforeseen problems with the building itself, contamination of the soil or an economic crisis. Frieden and Sagalyn (1990: 155) suspect that this way of sharing expenses often led to more public money being poured into the deal than was admitted. The need for entrepreneurial approaches was intensified by massive deindustrialisation of the urban economy as well as changes to taxation laws in the 1980s (Heeg 2008). In order to generate greater tax revenue, greater building activity was stimulated. 'To pay its bills Boston had little alternative but to permit more buildings and to increase the tax base' (Kennedy 1992: 214). In addition, a development-oriented logic was taken over and the city tried to calculate the profit margin of given projects and to tap into that profit. In order to offer an incentive for development and building projects, Chapter 121A on tax relief and greater zoning flexibility was frequently referred to. Corresponding rebates and incentives were negotiated with individual developers, that is, they were tailored mainly to specific projects and less to larger areas which were to be renovated as was the case previously up to the 1960s. In addition, these deals were about downtown. One result was hectic building activity, which, with delayed effect, was actualised in the 1980s.[10] Increased employment opportunities in sectors like the stock market, real estate, hotels and business services all contributed to a high degree of building activity with an increased demand for office space. In the 1980s the Boston skyline was created, with large sections arising out of speculative building projects. The shortage of developable plots encouraged more intensive exploitation of the space available by adding storeys and by expanding towards the South Boston Waterfront. By the end of the 1980s, public investment, along with the real-estate 'gold rush', had achieved a revaluation of the downtown area. Since the 1980s, demand for the product 'downtown' became so great that new and previously undiscussed areas came into focus.

---

10   Kennedy summarises the result of this policy as follows: 'In 1974 the city had only 14 million square feet of premium office space but by the end of 1984 it had in excess of 25 million square feet and much more was planned ... The assessed value of taxable Boston real estate increased tremendously through the decade after the court-ordered updating of the city's assessment practices in the early eighties. Total taxable value of $ 22.5 billion in 1987, and to $ 34.3 billion two years later. Revenues from property taxes account for about a third of Boston's city budget' (Kennedy 1992: 221–2).

Taken together, the projects which were carried out between the late 1970s and the end of the 1980s contributed to the rediscovery of the old downtown. The term rediscovery refers here not only to office buildings and tourist highlights (like Faneuil Hall Marketplace or the New England Aquarium), but also includes the residential building projects near the waterfront (Atlantic Avenue) as well as the transformation of the military port, Charlestown Navy Yard, into a more diversely used space. In general, downtown Boston was rediscovered as an interesting place for living, going out, engaging in recreation or communication. The newly created space is, however, a space which, despite many publicly available events, is coloured by strategies of commercialisation and monitoring.

The development of the South Boston Waterfront is also connected to an entrepreneurial approach to urban planning. The first spurts of development occurred along with the real-estate boom at the end of the 1980s. The leap to being the South Boston Waterfront, which occurred after the formerly elevated Interstate 93 was tunnelled so that only the Fort Point Channel separated it from the Financial Center, was a result of three projects: the Federal Reserve Bank of Boston at the junction of Atlantic Avenue and the South Boston Waterfront, the Federal Courtyard at the end of Fan Pier, and the World Trade Center. Contrary to the opinion that the attractive location would naturally lead developers to take an interest in the South Boston Waterfront, what can be observed is a very cyclical pattern of development which was influenced by demand for different types of real estate. Further, one must ask if the development plans constitute a new downtown or an expansion of the CBD as part of the old downtown. What makes the difference is the way that usage is distributed, that is, in what proportion residential usage stands to office space.

The South Boston Waterfront in Connection with the Old Downtown

The overall goal of an entrepreneurial urban planning policy in Boston was to establish the prerequisite infrastructure for a revaluation of the old downtown. This also applies to the South Boston Waterfront as the largest current development project in Boston. The opportunity to develop the South Boston Waterfront as a new downtown presented itself in the wake of the slow decline of the harbour starting in the 1950s. The possibility of actually carrying out development plans is, however, intertwined with public investments and real-estate cycles. This opens up a problematic context in which the interests of residents, city planners and representatives of the real-estate industry are not always compatible. The basic consensus since the end of the 1980s was that the South Boston Waterfront is the last piece of central property which could provide an opportunity to revitalise and reurbanise the city centre. This can be seen in the following quotation:

The waterfront, 1,000 acres stretching from the length of the Fort Point Channel to Massachusetts Bay, and now mostly a collection of vacant land, parking lots, and old warehouses, offers breathtaking views of the downtown skyline and proximity to the Financial District and Logan International Airport … But the South Boston waterfront has never been transformed into a bustling waterfront district such as those under way from Baltimore to San Francisco. Such developments are aimed at creating economic growth and luring tourists and suburbanites, who might be persuaded to move back to a revitalized city. City officials say that began to change last week when Mayor Thomas M. Menino and city planners unveiled a master plan for the South Boston waterfront. (Flint 1997)

The cause for this feeling of elation in the face of impending change was the publication of general guidelines in which the neglected harbour and industrial area were conceived of as a vibrant neighbourhood with things to do around the clock. This was to be achieved through the support of multiple public investment schemes starting in the mid-1980s, and which were to contribute to the attractiveness and value of the location. The most important measures were (a) the Harbor Clean-up Program, which improved the water quality, (b) the Central Artery, also known as the Big Dig, i.e. the depression of the Interstate 93 which connected the waterfront with the old downtown and improved access and (c) the Ted Williams Tunnel which connected the South Boston Waterfront to the highway system and to Logan International Airport. Since then, one can drive to Logan International Airport from the South Boston Waterfront in five minutes. Finally, (d) the Silver Line bus route, itself connected to the subway system, linked the waterfront to the city and the airport. These elements of the infrastructure and improvement of environmental conditions have all contributed to a substantial upgrade in the value of the location and an increase in real-estate value for waterfront property.

This raised expectations for real estate connected to the area. The South Boston Waterfront now has excellent infrastructure and locational advantages which have increased the attractivity of the area for housing and office space (and for retail space dependent upon resident and employee demand). There are many opportunities to use these advantages to create a space with unique qualities. At the end of the 1990s, speculation on the future development of the area reached its peak. Speculation is especially evident in the real-estate deals which led to a rise in property prices and also in the quantity of luxury apartments, most of which were aimed at the old downtown since those who owned property on the South Boston Waterfront wanted to keep it for office buildings (Heeg 2008).

In the period from 2000 to 2005, roughly 50% of the surface area of the South Boston Waterfront changed hands. In 2006, one of the last and most significant deals was made and a 24-acre property on the Inner Harbor was sold for 225 million dollars (Palmer Jr. 2006). Property on the Inner Harbor and especially those directly at the water's edge were considered the choicest cuts of the waterfront and it was assumed that that was where waterfront development would begin. The area had several advantages: a view of downtown, walking distance to the Financial District, access to water sports, easy access to the interstate and highway system and to the airport in just a few minutes, etc. Real-estate market processes, however, threw a wrench into these expectations. In 2008 and 2009, the areas in question seemed more likely to be the last to see construction, although the municipality willingly approved projects. The background to that was a constantly high proportion of unused office space since 2001 which, due to the financial crisis, was not a temporary slump but rather became the status quo. It became clear that the owners of the properties were not willing to start development with residential projects. Instead, they preferred to wait for a more opportune time for office real estate, with a view to supplementing office projects with a few apartments. As a result, development has been halting up to this day.

While caution, waiting and even complete stasis were the rule on the office market in the new decade, the market for apartments in downtown Boston could not keep up with demand, especially for high-end and luxury condominiums, and took off in 2004. The cause for this was a change in preference structures. Living in the city centre and on the water (or with a view of the water and downtown) was now a matter of high prestige. Prosperous households whose members are highly qualified workers in the service sector and baby boomers are considered to have a large market potential because of their interest in living in the city centre. Urban flair and access to diverse cultural and commercial goods, along with a high-quality and safe environment are all factors expected to attract both groups. However, the locations which can meet these demands are limited basically to Back Bay, South End, the South Boston Waterfront and other inner-city waterfront areas.

It is now possible to ask very high prices for downtown and waterfront housing (Figure 6.3). Only a very small segment of the apartments are affordable and can be paid for by households with average or below-average income.[11] A study by the real-estate agency Otis & Ahearn showed that the proportion of condominiums sold at under $500,000 decreased sharply between 2004 and 2005 (Blanton 2005).

---

11   Here one must take into consideration that the average regional household income is higher than the average city household income. Some high-end suburbs which are included in the metropolitan area inflate the average income. Basically, household incomes in the suburbs are higher than in the city (interview with neighbourhood activist Shirley Kressel, 8 June 2005 and with the former BRA policy director Tim McGourthy, 21 June 2005).

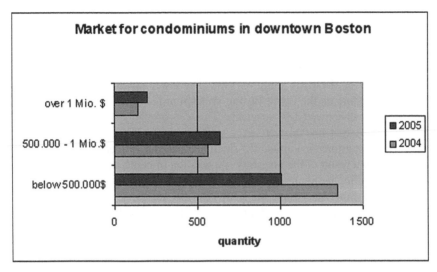

**Figure 6.3   The market for condominiums in downtown Boston (sales
between January and June each year)**

Note: Text should read: over 1 million USD, 500,000–1,000,000 USD, under 500,000 USD.
Source: Blanton (2005).

One indicator for the appeal of the inner-city real-estate market is that property
and real-estate funds started to include the purchase or building of downtown
residences in their portfolios (*Boston Business Journal* 2005). So while office-
only projects were treated sceptically, the opposite development could be observed
from the year 2000 on the market for condominiums and mixed-usage projects.
The headlines clearly manifest the reigning mood: 'Baby boomers to have their
say on luxury condo market' (*Boston Business Journal*, 15 October 2004), 'Hot
condo market leads developer to put apartments on block' (*Boston Globe*, 5 May
2005), 'Luxury condos more than double local sales pace. More than two dozen
condos traded hands for $2M or higher' (*Boston Business Journal*, 1 August
2005), 'Gold rush is on to build condominiums' (*Boston Business Journal*,
18 July 2005). Several condominium projects were built directly on the waterfront
in the Financial District, that is, on the way into the South Boston Waterfront
(Palmer Jr. 2005). Several properties which had been planned as office buildings
were changed into condominiums. This applied especially to building projects
along the entire downtown waterfront and the Rose Kennedy Greenway (what
used to be the Central Artery or Big Dig project (Palmer Jr. 2004)).

The South Boston Waterfront was to profit from the attraction of living
near the water because it was also close to downtown. While there were hardly
any new plans for office buildings in the area in the new decade, there were
several plans for apartment buildings. The majority were mixed-usage which
were re-integrated into planning because of the boom in the housing market.

The number of apartments which were built and planned, however, remained small in comparison with the old downtown. This has to do with the fact that the average floor space per apartment of 1,500 square feet was far more than the average demand projected in planning documents like the Seaport Public Realm Plan of 1999 or the BRA study of 1988. Far fewer apartments were actually built by investors on the South Boston Waterfront than had been envisioned by city planners or expected by residents. This can probably be explained by the fact that the target market were very wealthy buyers with a demand for very large luxury apartments. Further, the small number of apartments (Boston SAND 2009) is also a result of the investors' expectation of being able to turn a greater profit from office buildings. This expectation encouraged a cautious attitude on the South Boston Waterfront. Many owners did not carry out plans for office buildings when the market conditions were not favourable, but this did not mean that they constructed apartment buildings instead. Rather, they waited for market conditions to improve. Residents vehemently demanded that city planners put pressure on investors and developers to build apartments, if necessary by means of deterrents (e.g. higher taxes for undeveloped property). However, city planners and the BRA did not react to these demands. Up to this day, only a few projects have been completed, which can also be explained by the financial crisis. Since the summer of 2008, many projects have been put on ice, regardless of whether they were office or apartment buildings.

Ambitious plans for the development of the South Boston Waterfront as a new downtown have not come to fruition at the time of this writing. Office buildings were given a clear priority, but few of these plans have been realised due to unfavourable market conditions. On the other hand, the number of new residents has also remained small as a result of the type of apartments that have been built and the disproportion of office space. The liveliness which was a particular wish of both city planners and residents, and which was expected on the basis of the development of the old downtown, is not yet observable and it is not realistic to expect that to change in the near future.

Is the South Boston Waterfront a New Downtown?

Real-estate developers and city planners are aiming to capitalise on the advantages of the locations of the South Boston Waterfront, such as the proximity both to the water and to the CBD along with optimal access to traffic infrastructure.[12] In doing so, there is a tension between the desire to make it an independent space which is distinct from the CBD, but also to allow it to profit from the connection to the CBD. The ambitions of city planners are aimed at widely accepted ideas

---

12   This also applies to those few residents in the area who, since they do not own a large proportion of the property, do not have much influence. For an in-depth treatment of inverventions on the part of residents see Heeg (2008).

about high-quality living, shopping, restaurant and recreation standards. Here, 'widely accepted' means that the plans were also accepted by representatives of the real-estate sector, who also expect to achieve an increase in the value of the area by means of mixed usage. However, representatives of the real-estate sector are generally not open to discussion when the topic turns to how their own property should fit into the overall mixed usage plan. They are mostly open to a mix of parks, offices, apartments and museums in the new area to achieve distinct locations and diversity. It is expected that this mix will expand market potential and increase the value of each individual property. When it comes to their own property, they want to realise the highest possible value which the proximity to the CBD allows. Various projects have demonstrated thus far that there is not much openness to compromise when, for example, the proportion of office buildings is to be reduced to make room for apartments or more green space. Each property owner expects the others to be responsible for taking the steps needed to increase the value of the area overall, but refers to market forces when it comes to their own property (Heeg 2008). In sum, this attitude, which was not sanctioned by city planners, entails that those usages are realised which result from real-estate cycles and which tend towards monostructures. The general desire to create a unique space conflicts with the priorities of many investors which lead them to prefer to build offices, since that is what brings in the highest profit within the present situation. These aims, which until now have not been shared by the real-estate market, reduce the possibilities for creating a new downtown. A disproportionately large quantity of office space would make the South Boston Waterfront into a mere expansion of the CBD. The development of the South Boston Waterfront has proved to be a challenging balancing act between individual ambitions to make a profit and the overall coherent planning of the space.

Until 2009, more apartments than offices were built due to the booming real-estate market. This could not be foreseen at the highpoint of the market for offices at the end of the 1990s, but that does not mean that more apartments than offices will be built overall, but only that office projects are being delayed. The real-estate cycle does not only influence how quickly and whether or not construction is undertaken, but also what sorts of buildings are built. It seems the pulse of the real-estate market is the beat which developments on the waterfront follow.

It is clear that development happened in waves radiating from downtown. In times of economic growth, when there is a high demand for new offices and living space, the shortage of developable land and buildings that could be renovated led real-estate development to turn to the South Boston Waterfront. Whenever the market for offices and living space within the downtown area had been exhausted, development shifted over in the direction of the South Boston Waterfront. The fact that development moved in this particular direction and not into other adjacent areas such as the South End has to do with the attraction of waterfront property, but also with the circumstance that the other adjacent neighbourhoods are residential areas. Expanding commercial building into those areas would have led to significant conflict. On the other hand, the proportion of residents is very small in the South Boston Waterfront so there are still windows of opportunity in the question of usage.

Basically, this situation must lead to fundamental scepticism as to whether the South Boston Waterfront can be developed into a new downtown. Individual projects have been realised through planning and building projects, but they do not constitute a coherent urban space. The scarcity of apartments and, accordingly, of residents, especially brings up the question as to how a lively urban space is to be achieved. So far, the office projects lead one to expect that this neighbourhood will be inhabited during the day, but largely abandoned in the evenings. The question as to whether old and new downtowns augment or conflict with each other is hard to answer in this case. It is to be expected that primarily office buildings will be built in the near future. If this scenario becomes reality, then the two downtowns do not augment each other, but rather the new downtown is simply an extension of the CBD. On this scenario, the South Boston Waterfront will be, depending on the real-estate cycle, either a high-demand location or a B-location. In that case, the relationship between the old and new downtowns would be more contradictory. While the old downtown transitioned into having a relatively high proportion of living space with diverse opportunities for recreation in the last 20 or 30 years, the South Boston Waterfront is on the verge of becoming, if not a monostructure, then certainly an area of low usage variety. As such, the South Boston Waterfront cannot augment the downtown area, but can only become the poor step-child of the old downtown.

This is made possible because of urban development coming primarily out of the private sector (Squires 1996), in which real-estate investors are included in development less by means of the vision of urban planning, and more by means of private initiative as the basis for their development projects. When private projects do not correspond with public goals, the reaction is to provide incentives such as planning aides and financial support, but, in the case of the South Boston Waterfront, the effect of these has been neither that room was made for less high-profit usages nor for usages which brought in the highest possible profits. A lively and inhabited downtown, a downtown which is a place of encounter, which all parties say is their goal, threatens to remain an impossible vision. In Boston, city planning is understood to be an auxiliary task which is there to provide opportunities for private initiatives, but not to intervene with regulatory authority. Planning is focused on opening up scope for constituting the new area based on the idea that whatever private property owners do with that scope will be the optimal result for the urban community. This is certainly dubious, since, in expensive areas like the South Boston Waterfront, there is a strong tendency to prefer those usages which bring in the highest profit. A widely accepted vision of a lively neighbourhood without the willingness to intervene in private decision-making is like a tiger without teeth.

In sum, the proximity of the old downtown and the rediscovery of the city centre after years of decline are the preconditions for real-estate activity and development on the South Boston Waterfront. However, based on developments thus far, these factors have not encouraged the formation of an independent new downtown, but rather are leading towards an expansion of the Central Business District.

References

Blanton, K. 2005. Modestly Priced Condos Grow Rare. *Boston Globe*, 20 August.

Bluestone, B. and Stevenson, M.H. 2000. *The Boston Renaissance. Race, Space, and Economic Change in an American Metropolis*. New York: Russell Sage Foundation.

Boston Business Journal. 2005. Archstone-Smith Pays $1.1B for Oakwood Worldwide Apt. Sites. *Boston Business Journal*, 1 August.

Boston Municipal Research Bureau – BMRB. 2006. Boston's Property Values Trend Closer: Further Tax Shift to Residential Property Still Likely [Online]. *Special Report*, 06(2). Available at: www.bmrb.org/content/upload/report062. pdf [accessed: 3 January 2011].

Boston SAND – Seaport Alliance for a Neighborhood Design. 2009. *Report August 2006*. [Online]. Available at: www.seaportalliance.org/SAND/ Archive/050128housing.html [accessed: 12 October 2009].

Chapman, J., Facer, I. and Rex, L. 2005. *Connection Between Economic Development and Land Taxation. Land Lines*. Newsletter of the Lincoln Institute of Land Policy 4, 6–8.

Flint, A. 1997. Plan would Change S. Boston Waterfront: City's Proposal Calls for Hotels, Housing, Stores, and More to Draw Tourists and Residents to District. *Boston Globe*, 1 December.

Frieden, B.J. and Sagalyn, L.B. 1990. *Downtown, Inc.: How America Rebuilds Cities*. Cambridge, MA: The MIT Press.

Harvey, D. 1989. From Managerialism to Entrepreneurialism: The Transformation in Urban Governance in Late Capitalism. *Geografiska Annaler B* 71(1): 3–18.

Heeg, S. 2008. *Von Stadtplanung und Immobilienwirtschaft. Die 'South Boston Waterfront' als Beispiel für eine neue Strategie städtischer Baupolitik*. Bielefeld: Transcript-Verlag.

Horan, C. and Jonas, A. 1998. Governing Massachusetts: Uneven Development and Politics in Metropolitan Boston. *Economic Geography* 74(5): 83–95.

Jakle, J.A. and Wilson, D. 1992. *Derelict Landscapes. The Wasting of America's Built Environment*. Lanham: Rowman & Littlefield.

Kennedy, L.W. 1992. *Planning the City Upon a Hill. Boston Since 1630*. Amherst: University of Massachusetts Press.

Keyes, L.C. 1969. *The Rehabilitation Planning Game: A Study in the Diversity of Neighborhood*. Cambridge, MA: The MIT Press.

McQuade, W. 1966. Urban Renewal in Boston, in *Urban Renewal: The Record and the Controversy*, edited by J.Q. Wilson. Cambridge, MA: The MIT Press, 259–76.

New Urbanism

Mollenkopf, J.H. 1983. *The Contested City*. Princeton: Princeton University Press.

O'Connor, T.J. 1993. *Building a New Boston. Politics and Urban Renewal 1950–1970*. Boston: Northeastern University Press.

Pagano, M.A. and Bowman, A. 1995. *Cityscapes and Capital. The Politics of Urban Development*. Baltimore and London: Johns Hopkins University Press.

Palmer Jr., T.C. 2004. For Property Owners, Parks mean Profits. *Boston Globe*, 14 June.

Palmer Jr., T.C. 2005. Big Hotel to be Cut from Plan for Wharf. *Boston Globe*, 8 July.

Palmer Jr., T.C. 2006. McCourt Reportedly Selling S. Boston Site. *Boston Globe*, 14 January.

Sagalyn, L.B. 1989. Measuring Financial Returns. When the City Acts as an Investor. Boston and Faneuil Hall Marketplace. *Real Estate Issues* 14: 7–15.

Squires, G.D. 1996. Partnership and the Pursuit of the Private City, in *Readings in Urban Theory,* edited by S.S. Fainstein and S. Campbell. Cambridge: Blackwell, 266–90.

Walker, A. 1999. Today's Boston is His Legacy. *Boston Globe*, 29 July.

Ward, S.V. 2002. *Planning the Twentieth-Century City. The Advanced Capitalist World*. Chicester: Wiley.

Weber, R. 2002. Extracting Value from the City: Neoliberalism and Urban Redevelopment, in *Spaces of Neoliberalism. Urban Restructuring in North America and Western Europe*, edited by T. Brenner and N. Theodore. Malden: Blackwell, 172–93.

Wikipedia 2009a. *Government Center in Boston* [Online]. Available at: http://en.wikipedia.org/wiki/File:Government_Center_Boston_vista.jpg [accessed: 17 November 2009].

Wikipedia 2009b. *Quincy Market in Boston, Massachusetts* [Online]. Available at: http://en.wikipedia.org/wiki/File:Quincy_Market_1.JPG [accessed: 17 November 2009].

Chapter 7

Grasping, Creating and Commercialising Trends, Styles and 'Zeitgeist': The Role of Urbanity with Regard to Working in Flexible, Specialised Project Networks as Illustrated by the Media Industry

Ivo Mossig

Introduction and Purpose

Urban economies and creative growth industries such as the media industry have increasingly gained the attention of development plans centred around regional economies. Many cities have undertaken rigorous efforts in developing their own location to become a media cluster in order to compensate for the structural transformation of the associated process of deindustrialisation. A conspicuous feature from a geographic perspective is the high spatial concentration of the media industry in the urban centres. According to information from the Federal Agency of Labour for 2007, 799,556 people were in gainful employment requiring mandatory contributions to national insurance in Germany's four largest media industries: (a) printing and publishing, (b) software manufacturing, (c) advertising and (d) radio and television. Accordingly, these four areas of the media industry in Germany account for 3% of all jobs requiring mandatory national insurance contributions. Munich occupies the top slot of the four most important media industries with 48,899 employees, followed by Berlin (46,780) and Hamburg (46,414). These three cities, which are at the same time Germany's largest urban centres, therefore form the alpha category of media locations in Germany. In the beta category of German media clusters, we find the next largest cities Cologne (30,307 media employees in the four aforementioned areas), Frankfurt (22,462) and Stuttgart (21,728) (Mossig and Dorenkamp 2008).

The purpose of this study is to show that creative growth industries such as the media industry require a lively urban environment due to their characteristics and the main thrust of their operations. This is intended to explain why the media industry and also other areas of the creative industries are concentrated on the largest cities and metropolis regions. Furthermore, it shall be indicated that a highly flexible organisation of the production system and the division of labour between companies in the form of primarily local project networks (Manning and Sydow 2005; Mossig 2004; Sydow and Staber 2002; Windeler et al. 2000) have contributed decisively to the success of the media industry. Drawing on the example of the production of film and television broadcasts, the intention is to show that project networks open up the individual freedom for the parties involved to develop the most original and creative ideas possible, to draw on the individual skills for the entire project and to ensure a prompt economic pay-off. The interesting factor in this is that working in projects is also becoming increasingly important in other areas of the economy. Accordingly, the so-called cultural product industries, i.e. the creative industries (Mossig 2005, 2006; Power 2003; Scott 2000), among which the media industry is always numbered, are considered to play a pioneering role. This means that the example of the media industry can be used to analyse current established practices of labour organisation in project networks that most probably will permeate more and more knowledge-centric areas of work (Krätke 2002).

The study is structured as follows: the first section will provide a brief summary of the main characteristics of the media industry. What distinguishes the media from other industries and why is an attractive urban environment of special importance for the media industry? The next section will deal with the defining features, the coordination mechanisms and executive structures of the flexible project networks within the media industry. The intention in this is to establish that the organisation with its pronounced division of labour within local project networks also reaps the rewards of a lively, diverse, open and communicative urban environment.

Up-to-Date and Current – The Characteristics of the Media Industry

It was not until the last few years that the media industry entered into the focus of the research work on economic geography and the regional economic development concepts. The positive growth rates since the 1980s should be mentioned as a significant reason for this new recognition (compare Figure 7.1).

The core business of the media industry is to generate content, this being information, messages or cultural commodities such as music or films, and to distribute them via its own platforms. In this, the media are assigned the task to fulfil an independent control function with regard to state and other public sector institutions and to mediate between the complex social and political spheres by selecting, preparing and publicly distributing information (Beck 2002). These days, however, the field of entertainment content distributed via the media enjoys an increasingly greater significance, as profits can be generated if customers are willing to focus attention on the information and content produced.

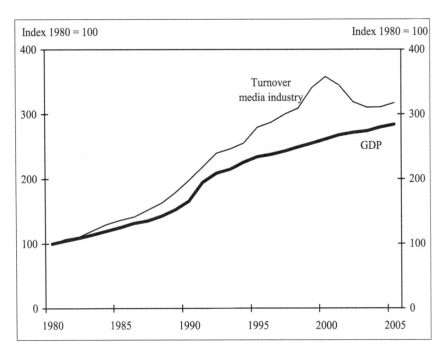

Figure 7.1   Development of turnover in the media industry in Germany in comparison with the gross democratic product (GDP) as indicator for the overall economic development 1980–2005

Source: Statistical Almanac.

'We pay attention', we pay with attention in the most literal sense of the word. But it is less the quality of the content and more the print run of a print medium, the viewer figures for a television broadcast or the click on an Internet site that determine the commercial success of a media company (Beck 2002; Mathes et al. 2001). Therefore, we are dealing with attention economies (Franck 1998), i.e. with economies of fascination (Schmid 2006).

In order to attract the greatest possible attention, it is of prime importance for the protagonists within the media industry to operate in a way that is as up-to-date and current as possible. They are constantly called upon to generate a stream of new emotionalising contents that mirror the consumers' lifestyles or that have the potency to play a defining role in them. This is why those involved in the media industry must draw on and exploit new trends and fast-paced manifestations of the zeitgeist in order to create the most original, creative and, above all, attention-grabbing content. In so doing, the media not only draw on the latest trends and styles, they also play an active role in establishing them, e.g. through the presentation of trendsetters. Localisation of the media industry in the urban centres can be explained on the basis of the significant importance to the protagonists of being up-to-date and current. In addition to the direct proximity to the nerve centres of the political and economic spheres as a source of content suitable for the media, the greater internationality, tolerance and openness (Florida 2002) of the urban centres in comparison to the more peripheral locations offers considerably more leeway to try out and embrace the latest fashions, looks and individual lifestyles. This means we find a special diversity in the cities and therefore a special place to receive inspiring ideas. In addition, within an urban context, trends, fashions and lifestyles require a special quality in order to rise above the many different trendsetters and protagonists of a modern lifestyle and therefore to secure suitable attention and respect. Consequently, to a greater degree, the latest manifestations of the zeitgeist are found in their most manifold variety and most distinct presentation forms in the urban centres. Urbanity therefore enables employees in creative areas of the media industry to perceive the latest trends and styles at all times, to draw inspiration from them and to transform these impressions and experiences into commercially exploitable media content as well as to try out their own creations on the ground. In Germany also, this leads to the high concentration of activity within the media industry observed in the largest cities.

It is important, furthermore, to identify the significance of creative processes that take place within the framework of creating media products. Creativity represents a fundamental driving force during innovative processes and is therefore an important engine of economic developments that until now have received paltry attention as a production factor. Although the term 'creativity' is drawn on in work relating to the 'creative milieu', it is viewed less as an individual personality characteristic, but more with regard to collective learning processes resulting from an intense networking between the protagonists within the milieu (Becattini 1991; Fromhold-Eisebith 1995). Accordingly, Franz (1999) speaks conclusively of 'innovative milieus'. Creativity is an individual characteristic that, unlike the specialised expertise of highly-qualified purveyors of knowledge, is not based on a university education. Hence, compared with knowledge, it is considerably harder to acquire creativity through learning processes, if it can be acquired at all in the form of a targeted learning process. This is mirrored, for instance, in the fact that a disproportionately large number of professional persons from other industries without corresponding, specialist education occupy the highest positions within the media industry with great success. Creativity can hardly be substituted; it can neither be purchased nor activated at the drop of a hat, nor can it be carried in a box from one office to the next. Creative thought and action are characterised by innovation and originality, and as such, may certainly break rules and subvert or even constitute an act of rebellion. Consequently, creativity is more than a simple process of alignment, improvement or optimisation; it is instead a force that brings forth new knowledge and 'innovative alternatives' (Hauschildt 1997: 301; Mossig 2006). Creative people very rarely act within an isolated context or essentially removed from their environment, even if the perception of a composer at his piano leading the life of a hermit or of an author sitting in the garden of his rural country retreat writing his texts on a laptop falsely appears to suggest this independence. Quite to the contrary, creative processes are generally spawned by a creative person's extremely intense critical appraisal of his or her environment (Törnqvist 2004). In view of the fact that structured education such as university studies can only promote creative processes to a limited degree, but at the same time a diverse, exciting, open and tolerant environment has been recognised as a source of inspiration for creative protagonists, it is fair to derive in total the need to cultivate an environment that fosters creativity to the greatest possible extent and in which diversity, once deemed the beacon of economic inefficiency, becomes an important resource. Urban locations in particular can offer these advantages of diversity (Grabher 1994; Helbrecht 1998, 2005; Mossig 2006; Thiel 2005; Törnqvist 2004).

Thus, statements made by the experts from the media industry when asked about the influence of urbanity in Berlin on work within their industry emphasise the significance of an open, diverse, exciting and lively environment that is up-to-date and current (compare Mossig and Dorenkamp 2008):

Berlin is a constant challenge ... The city also offers the opportunity of an entirely unconventional approach and the ability to transform everything of interest, everything that interests me, into money or products, as there is space for all that here. (TV producer, Berlin)

By nature, Berlin is ... far more dynamic. There is more happening, life is simply faster ... And I also worked in Baden-Baden, and it is really bad there. It is like a sheltered life there. But Berlin is different ... That kind of sheltered life is deadly for my profession. In this instance ... an urban environment like that really is important. I certainly like to immerse myself deliberately in the kind of social situation you find in Berlin. (Scriptwriter, Berlin)

I believe you can find your feet far easier in Berlin due to the more open system and the more cosmopolitan community here with its over 3 million people. (TV producer, Berlin)

Actors love living in the trendiest, most exciting centres as we currently find in Berlin. (Film producer, Munich)

Berlin is something special ... It is the only city that can offer 3.5 – 4 million inhabitants. International business senses this expansiveness. There is certainly an interesting subculture here and that is why there is a tractor beam that always attracts the media. (Studio director, Berlin)

Division of Labour between Companies, Local Project Networks and Urbanity

However, the spatial concentration on the urban centres is not solely due to the advantage explained above that the protagonists are able to operate in a particularly inspiring environment there. Ultimately, skills inherent to persons such as individual knowledge levels, memory, subjective experience and the consequent assessments thereof, along with their creativity, are insufficient to create and successfully market a complex product. The intellectual capacities of one individual are soon exceeded in an increasingly complex information society with a rapidly rising reservoir of knowledge. This is also the reason why organisation in almost all areas has a distinct division of labour. Structures that draw on the division of labour enable parallel absorption of new information via several channels of information and therefore a vast enlargement of possible contacts for new inspiration and ideas. Information processing also takes place much faster in a structure with division of labour than any individual person could achieve. Consequently, more knowledge and information can equally be processed and used efficiently by coordinating the actions of many individual persons than could possibly be the case with an identical number of uncoordinated individual protagonists (Meusburger 2007). However, the coordination of processes with division of labour is not free of charge. The transaction costs themselves differ, depending on the frequency and specificity of the associated transactions (Williamson 1990). In this respect, Scott (2006) emphasises that individual interactions can be differentiated in terms of the spatial dimension. Spatial proximity is of far greater significance for some interactional relationships than for others; the spatial transaction costs differ in their amounts accordingly. This explains why some economic activities are concentrated on one location and form a cluster due to their high spatial transaction costs, whereas other areas see a dispersed spatial distribution pattern as a result of the lower spatial transaction costs (Scott 2006).

In addition to the general benefits of division of labour between companies, a flexible organisation of the production system is necessary in order to achieve the fastest possible commercial exploitation in the rapidly moving appearance of new trends, styles or the zeitgeist. Additionally, a further requirement for the production system is to enable individual space for the original implementation of new, creative ideas, while at the same time ensuring rapid response to new requirements. This flexibility and the required space are equally achieved in the media industry through a suitably high degree of division of labour between companies. The value added system is almost entirely dissected and each of the tasks is outsourced to individual, independent companies and freelancers. The individual companies and the freelancers come together for the duration of processing (the joint project). The network for the production of the media content disbands once more upon completion of the project task. The affiliated firms and freelancers then usually work on the following projects in a different constellation that matches the new project requirements. Project networks of this kind (Manning and Sydow 2005; Mossig 2004; Sydow and Staber 2002; Windeler et al. 2000) are extremely flexible as the individual firms have built up specialised know-how in their field of operation over time due to their focus on one partial area of the value added system and are therefore able to respond flexibly to new requirements in their field and to develop creative solutions. In other words: the flexibility is derived from individual specialisation benefits and the composition of the project network with specialists to meet the requirements of the project (Mossig 2004).

There is also an extremely high degree of flexibility offered to project initiators. Problems of capacity utilisation and overheads are transferred to the individual subcontractors. Additionally, colleagues in this kind of project network are highly motivated, as follow-up contracts are awarded above all to those colleagues who were able to draw attention to themselves in the local media scene due to good performance in previous projects. Helen Blair succinctly entitled this phenomenon with the words 'You're only as good as your last job' (2001). Furthermore, a project-based cooperative of many individual firms provides the freedom required in order to exploit the individual creative potential and specialisation benefits of each firm and freelancer as effectively as possible.

Let us take the example of the television industry in order to take a look at how this kind of project network forms (compare Figure 7.2) and which executive powers determine the intercompany coordination within the clusters (compare below Mossig 2004). During the run-up phase of a TV production, an independent production firm, the potential contractor, negotiates with a TV station about the idea for a broadcast, e.g. a TV movie. The production firm had previously developed the idea for the film with an independent scriptwriter. The station singles out from the pool of many ideas presented the one it believes will interest the greatest TV audience.

The details are clarified over the course of intense negotiations between the station and the producer, after which the station agrees to finance the TV movie project. In the next stage, the producer selects the key positions, the so-called heads of department. These are the director, who has contributory rights in selecting the lead actors, and the head of department for the equipment, as well as the production director and the studio as technical service provider. As contractor and financier, the TV station is entitled to veto any recruitments. After this, the heads of department put together their teams from the pool of independent firms and freelancers; the producer then pays them for the duration of their cooperation in the project using the budget previously negotiated with the TV station. Accordingly, the director selects his assistant and cameraman. The cameraman determines in turn his camera assistants. The head of department for equipment selects his makeup artists, stage designers and costume designers, etc., and so forth. This means that the respective specialists can be appointed flexibly to the individual positions, depending on the genre of the film. Whenever necessary, a different project team will be put together for a romantic comedy than for a tense action thriller.

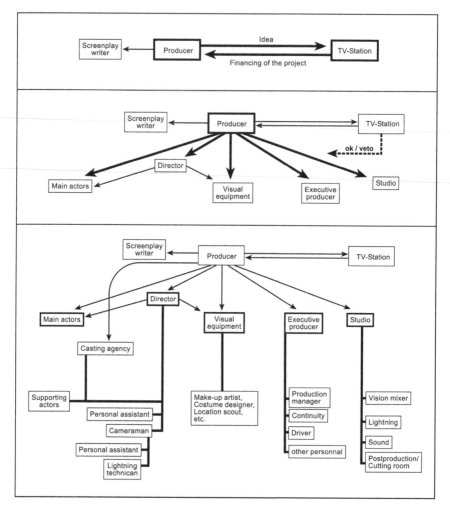

**Figure 7.2   Composition of the project network for the production of a
              TV movie**

Source: Mossig (2006: 122).

    Selection of the staff participating in a project is a truly decisive moment
with regard to coordinating the project networks (Windeler et al. 2000). Within
this context, however, the executive powers are spread in an extremely unequal
manner, meaning that defining the clear responsibilities is among the principal
control problems within local project networks (Mossig 2004; Staber 2000).
In general, the TV station is the most powerful protagonist, selecting as it
does the producer in an initial stage. The station has the money to finance the
project, and also the platform for marketing the TV broadcast in the media.

Consequently, the station has an upstream power position, as it draws on the two resources of money and broadcasting slot. However, project networks are not solely coordinated via the power inherent to holding resources (Bathelt and Taylor 2002; Mossig 2004; Taylor 2000). If one analyses the criteria in the selection of project staff, it becomes clear that the networks are also controlled by the power and influence of relationships. The most important positions in particular tend to be awarded to individuals who have established a certain reputation. For instance, an experienced director has more influence and contributory rights in occupying positions than an unknown novice. The project networks experience additional control through collective regulatory powers. News of any gross misbehaviour within a project spreads like wildfire in the media industry; in future, therefore, the protagonist in question will not only be shunned by the current contractor, but also by other persons initiating a project. This means that no protagonist can get away with opportunistic behaviour for any length of time. The informal networks within which this kind of information on the reliability and trustworthiness of individual firms and freelancers are therefore a simple and effective control and guarantee system (Bathelt and Taylor 2002; Manning and Sydow 2005; Mossig 2004, 2006; Taylor 2000).

An interview partner hit the nail on the head in this respect as follows:

> I talk to the Heads of Entertainment at other TV stations. We do swap stories about our experience with producers, and we do rely on the information, even if it does come from the competition. Warnings about bad experiences count especially. (Head of the Entertainment Department at a TV station, Munich)

Another discussion partner emphasised:

> In the media industry, we not only talk a lot with each other, but also a lot about each other ... For example, people say about one company: every second person who works with them ends up in court. (TV studio operator, Munich)

Therefore, in view of the fact that it is not just the available resources (e.g. in the form of funding) that coordinate the work within a project network, but equally also that influence is exerted through relationships and that collective regulatory powers show their parallel effects, personal contact networks and an informal flow of information are of particular significance. Face-to-face contacts in particular play a large role in this respect as the most efficient and intensive form of networking information. After all, body language and complementary gestures transport further information beyond the spoken word, thus contributing to a more simple understanding of the contents expressed. Recipients of the message also find it easier to recognise irony or emphasis on particularly relevant aspects in face-to-face meetings than could be the case in a telephone conversation or in written emails (Storper and Venables 2004).

Consequently, the project networks that actually form in order to work on a project contract are superseded by the individual persons' information networks, within which personal communication plays a weighty role. As we have seen, these individual information networks are of pivotal importance for coordinating the actual project networks.

Spatial proximity between the protagonists and cluster structures within a diverse urban environment foster the work in local project networks. Firstly, this organisational form requires a large pool of differently specialised firms, freelancers and creative individuals, so that general urbanity benefits (Gaebe 2004) come into play. Secondly, an attractive, inspiring workplace has a more enticing effect on the creative protagonists than a boring environment. Thirdly, the spatial proximity promotes establishment of the important information networks due to the opportunity to meet different people from the media scene in so-called neutral locations (i.e. outside of the purely private or professional radius of action). It is important for the individual protagonists to be on the ground in the local media clusters in order to gain access to the informal flow of information and in this way to enter into consideration in putting together the actual project networks or to find the right partners in the event of selecting the protagonists involved. Fourthly, the spatial transaction costs are comparatively high as a result of the significant need for personal networking of information and communication, and can be reduced accordingly if there is spatial proximity between the protagonists. Conversely, this means that, due to the significant need for insider information, the spatial concentration of creative and also knowledge-intensive growth industries can be linked to the spatial transaction costs.

The special significance of communication and the informal flow of information is emphasised as follows in three interview quotations from the field of the media industry:

> The personal contacts open the door, without which you stay out in the cold without a chance to seal any deal. (Actor's agent, Munich)

> Ultimately, there is a huge need for communication in order to complete a project the way you imagined it. (Content Manager at a TV station, Cologne)

> If I had tried to build up a company in Bottrop-Kirchhellen with access to all kinds of capacities, I would have ended up dead. I would not have been able to ... draw on a private network. And I am absolutely convinced I would have noticed that lack in commercial terms also. (TV studio operator, Munich)

Summary and Outlook

In essence, this chapter was intended to emphasise the mutual interaction arising from the flexible production system within project networks and the urban location requirements for creative growth industries such as the media industry. On the one hand, an urban environment represents a source of inspiration for protagonists in the creative areas of the media industry in order to absorb new trends, styles and the fast-paced manifestations of zeitgeist and to transform them quickly into fascinating contents. In view of the fact that in general, creative processes do not take place hermetically, but instead find their source as part of intense environmental scrutiny by the creative individual, urbanity offers a far greater wealth and width of inspiration compared with more peripheral locations. On the other hand, the variety on offer in an urban location represents an important precondition in order to establish a flexible production system in local project networks, which, in addition to the general benefits found in division of labour, permits each protagonist a high degree of individual freedom. Reference has been made against this backdrop to the fact that in Germany also, the media clusters are located in the country's largest cities, with Munich as the most important centre, followed by Berlin and Hamburg.

Spatial proximity facilitates communication between the protagonists. Personal contacts (face-to-face) in particular are easier to organise. This refers to both communication within an actual project network, in which coordination of the division of labour between the participating persons and companies is necessary, and also to the informal networking of information in order to establish and maintain individual contact networks. However, personal contacts require actual locations in which the meetings can take place. Business meetings are scheduled in advance, and a definite venue is agreed. Thus, the business partners can agree in advance on a location that appears appropriate for the business appointment in question. Conversely, the manner in which individual contact networks form and solidify takes place with a far lower degree of planning. Coincidental encounters especially, which take place in some cases outside of actual working hours, are very significant in this respect. In this, certain clubs, bars or events specifically within the media industry such as premiere parties, along with diverse and colourful happenings or hip locations from the world of so-called subculture, play an important role as more or less coincidental meeting places for the industry. Urban districts home to many small businesses offering original wares, inimitable bars, cafés and restaurants in an architecturally appealing environment that rises above the standards of humdrum development, are said to offer a particular quality conducive to lingering. It therefore appears sensible to provide corresponding space in order to develop individual, small, diverse and autonomous activities, i.e. to keep this space open and affordable, thus ensuring the emergence of a mixture that appears attractive to creative individuals. The objective in this should be to create and sustainably establish a so-called place for communication, inspiration and personal information networking.

In addition, the contents herein have clearly shown that the power relationships and executive powers are divided unequally between the protagonists in project networks with division of labour. It is therefore very important for the development of any location that the powerful aspects of the value added system in particular are sourced locally. The powerful executives' radii of action represent the point of attraction for the many small, specialised firms and freelancers searching for the next temporary project collaboration. If there are no relevant contractors on site, the opportunities for a knowledge-intensive or creative growth industry to develop will remain restricted in any location.

References

Bathelt, H. and Taylor, M. 2002. Clusters, Power and Place: Inequality and Local Growth in Time-Space. *Geografiska Annaler B* 84(2): 93–109.

Becattini, G. 1991. The Industrial District as a Creative Milieu, in *Industrial Change and Regional Development: The Transformation of Industrial Spaces*, edited by G. Benko and M. Dunford. London and New York: Belhaven Press, 102–14.

Beck, H. 2002. *Medienökonomie. Print, Fernsehen und Multimedia*. Berlin, Heidelberg and New York: Springer.

Blair, H. 2001. 'You're Only as Good as Your Last Job': The Labour Process and Labour Market in the British Film Industry. *Work, Employment & Society* 15: 149–69.

Florida, R. 2002. *The Rise of the Creative Class: And How it's Transforming Work, Leisure, Community and Everyday Life*. New York: Basic Books.

Franck, G. 1998. *Ökonomie der Aufmerksamkeit*. München: Hanser.

Franz, P. 1999. Innovative Milieus: Extrempunkte der Interpenetration von Wirtschafts- und Wissenschaftssystemen. *Jahrbuch für Regionalwissenschaften* 19: 107–30.

Fromhold-Eisebith, M. 1995. Das 'kreative Milieu' als Motor regionalwirtschaftlicher Entwicklung. *Geographische Zeitschrift* 83(1): 30–47.

Gaebe, W. 2004. *Urbane Räume*. Stuttgart: Ulmer.

Grabher, G. 1994. *Lob der Verschwendung. Redundanz in der Regionalentwicklung. Ein sozioökonomisches Plädoyer*. Berlin: Edition Sigma.

Hauschildt, J. 1997. *Innovationsmanagement*. München: Vahlen.

Helbrecht, I. 1998. The Creative Metropolis. Services, Symbols and Spaces. *International Journal of Architectural Theory* [Online] 3(1). Available at: http://tu-cottbus.de/theoriederarchitektur/wolke/X-positionen/Helbrecht/helbrecht.html [accessed: 6 January 2011].

Helbrecht, I. 2005. Geographisches Kapital – Das Fundament der kreativen Metropolis, in *Knoten im Netz. Zur neuen Rolle der Metropolregionen in der Dienstleistungswirtschaft und Wissensökonomie*, edited by H.J. Kujath. Münster: LIT-Verlag, 121–55.

Krätke, S. 2002. *Medienstadt. Urbane Cluster und globale Zentren der Kulturproduktion.* Opladen: Leske + Budrich.

Manning, S. and Sydow, J. 2005. Arbeitskräftebindung in Projektnetzwerken der Fernsehfilmproduktion. Die Rolle von Vertrauen, Reputation und Interdependenz, in *Entfesselte Arbeit – neue Bindungen. Grenzen der Entgrenzung in der Medien- und Kulturökonomie*, edited by N. Mayer-Ahuja and H. Wolf. Berlin: Edition Sigma, 185–219.

Mathes, R., Möller, A. and Hißnauer, C. 2001. Medienerfolg durch Medien-Hype. Wie im zunehmenden Wettbewerb um die Aufmerksamkeit des Publikums die selbstreferentiellen Mechanismen des Mediensystems an Bedeutung gewinnen, in *Hundert Tage Aufmerksamkeit. Das Zusammenspiel von Medien, Menschen und Märkten bei 'Big Brother'*, edited by K. Böhme-Dürr and T. Sudholdt. Konstanz: UKV-Verlag, 63–77.

Meusburger, P. 2007. Organisation und Akteure in der Wirtschaftsgeographie, in *Geographie: Physische Geographie und Humangeographie*, edited by H. Gebhardt et al. München and Heidelberg: Spektrum Akademischer Verlag, 677.

Mossig, I. 2004. Steuerung lokalisierter Projektnetzwerke am Beispiel der Produktion von TV-Sendungen in den Medienclustern München und Köln. *Erdkunde* 58(3): 252–68.

Mossig, I. 2005. Die Branchen der Kulturökonomie als Untersuchungsgegenstand der Wirtschaftsgeographie. *Zeitschrift für Wirtschaftsgeographie* 49(2): 97–110.

Mossig, I. 2006. *Netzwerke der Kulturökonomie. Lokale Knoten und globale Verflechtungen der Film- und Fernsehindustrie in Deutschland und den USA.* Bielefeld: Transcript Verlag.

Mossig, I. and Dorenkamp, A. 2008. Mediencluster und die Bedeutung Berlins als Medienstandort. *Arbeitsberichte des Geographischen Instituts der Humboldt-Universität zu Berlin* 142: 55–73.

Power, D. 2003. The Nordic 'Cultural Industries': A Cross-National Assessment of the Cultural Industries in Denmark, Finland, Norway and Sweden. *Geografiska Annaler B* 85(3): 167–80.

Schmid, H. 2006. Economy of Fascination: Dubai and Las Vegas as Examples of a Thematic Production of Urban Landscapes. *Erdkunde* 60(4): 346–61.

Scott, A.J. 2000. *The Cultural Economy of Cities. Essays on the Geography of Image-Producing Industries.* London, Thousand Oaks and New Delhi: Sage.

Scott, A.J. 2006. *Geography and Economy.* Oxford: Clarendon Press.

Staber, U. 2000. Steuerung von Unternehmensnetzwerken: Organisationstheoretische Perspektiven und soziale Mechanismen, in *Steuerung von Netzwerken. Konzepte und Praktiken*, edited by J. Sydow and A. Windeler. Opladen: Westdeutscher Verlag, 58–87.

Storper, M. and Venables, A.J. 2004. Buzz: Face-to-Face Contact and the Urban Economy. *Journal of Economic Geography* 4: 351–70.

Sydow, J. and Staber, U. 2002. The Institutional Embeddedness of Project Networks: The Case of Content Production in German Television. *Regional Studies* 36(3): 215–27.

Taylor, M. 2000. Enterprise, Power and Embeddedness: An Empirical Exploration, in *The Networked Firm in a Global World. Small Firms in New Environments*, edited by M. Taylor and E. Vatne. Aldershot: Ashgate, 199–233.

Thiel, J. 2005. *Creativity and Space: Labour and the Restructuring of the German Advertising Industry*. Aldershot: Ashgate.

Törnqvist, G. 2004. Creativity in Time and Space. *Geografiska Annaler B* 86(4): 227–43.

Williamson, O.E. 1990. *Die ökonomischen Institutionen des Kapitalismus. Unternehmen, Märkte, Kooperationen*. Tübingen: Mohr.

Windeler, A., Lutz, A. and Wirth, C. 2000. Netzwerksteuerung durch Selektion. Die Produktion von Fernsehserien in Projektnetzwerken, in *Steuerung von Netzwerken. Konzepte und Praktiken*, edited by J. Sydow and A. Windeler. Opladen: Westdeutscher Verlag, 178–205.

Chapter 8

Major Town Planning Projects in Urban Renaissance: Structuring Property Sales for Future Urbanity?

Maike Dziomba

Major urban renaissance[1] projects such as Hamburg's HafenCity are experimental laboratories, even standard-bearers for planned urbanity. However, how can municipal government and the real-estate sector ensure that the new districts and neighbourhoods ultimately appear as almost naturally evolved, displaying the desired urban variety and developing a vibrant city life?

Against the backdrop of this question, this chapter uses the example of HafenCity to focus more on the phase of property sales in sub-projects than on the factual feasibility of urbanity as a result of planning. Which strategies in the real-estate sector can be exploited to achieve the planning objectives, among others the creation of urbanity? After all, the public sector, as bargainer, represented in Hamburg's case by the highly-professional quango,[2] HafenCiy Hamburg GmbH, certainly does have ways and means of securing at least some aspects of future urbanity: mixed uses with both a high residential quota and also 'attractions' with high public appeal, varied architecture and urban development designed with the needs of the people in mind, inviting public spaces, social heterogeneity and not least a certain variety in the ranks of protagonists involved (Dziomba 2009).

---

1   The term 'Urban Renaissance' originates from England, where it became established in 1999 as a 'new vision for urban revival' (compare Bodenschatz 2005; Urban Task Force 1999, 2005).

2   The acronym qua(n)go stands for 'quasi (non-)governmental organisation'; Schubert and Klein (2006) explain: 'Increasing numbers of politically active organisations can no longer be assigned clearly to the public or the private sector; in this, they respond to the often fluid transition between governmental (administrative) organisations and private (corporate) organisations found frequently in political practice. In terms of legal form, quangos are non-governmental organisations, which, in factual terms, fulfil state tasks, i.e. are widely influenced by state institutions'.

Caught between the priorities of the public sector and the property market, the central question is therefore: how can the many protagonists who are from the private sector – and therefore profit-centric – be drawn into a commitment for additional concessions and measures relating to planning-policy visions of an urban, attractive district? What means of governance does the bargainer have in this respect during the phase of property sales?

A summarised answer concerning the motives harboured by the private-sector protagonists (Dobberstein 2000) will boil down to short-term developer profits, long-term appreciation of the property or optimising conditions for personal use – the creation of urbanity could therefore be a mutual goal. Nevertheless, the protagonists are only willing to accept the more elaborate procedures and measures if they are equally binding and transparent for all those involved, their impact is predictable and all aspects of the major project proceed successfully.

This chapter uses HafenCity as an example in order to demonstrate how inherent preconditions for urbanity can be achieved within the framework of property sales as well as in the context of urban governance and the property market. Firstly, though, some basic conditions of central importance are explained.

Qualitative-explorative Study of Major Urban Renaissance Projects

The statements made here are based on a comparative, empirical study completed in 2007 on strategies of property development within the framework of major urban renaissance[3] projects. HafenCity in Hamburg and Westhafen in Frankfurt, which unlike the quango model in Hamburg is a PPP, were evaluated qualitatively on the basis of case studies and interviews with experts.

In this, the research interest required a theoretical framework of analysis that integrated the property sector perspective in the planning process. Drawing on a backdrop of governance theory, the questions were analysed on the interface between major project research, the governance approach and project development in the property sector.

In methodological terms, the study was rooted in fact-based theory construction, i.e. 'grounded theory' (Glaser and Strauss 2005). A characteristic feature of this is that partially at least, compilation, evaluation and analysis run parallel, whereby they mutually influence and support each other.

---

3 The main contents and statements in this chapter are based on the author's dissertation, submitted in 2007 and published in 2009 (Dziomba 2009).

The results of the study refer to the project organisation and governance instruments in major urban development projects of urban renaissance. Against the backdrop of the risks inherent to the property sector, the respectively suitable sales procedures are assigned to the different property types in accordance with the progress of the project.

Brief Presentation of HafenCity Hamburg

The City of Hamburg is developing the major project HafenCity on former port areas. The development firm GHS, Gesellschaft für Hafen- und Standortentwicklung GmbH, was founded as a city-owned company in order to design and implement the project. It was renamed HafenCity Hamburg GmbH in February 2004. The project magnitude and the timeline (see Table 8.1) indicate the project's significance for the city centre of Hamburg.[4]

Table 8.1   General information on the major project HafenCity Hamburg

Project magnitude, timeline	155 ha total surface, of which roughly 55 ha water; project initiation in 1997, duration at least 25 years; expansion of the city centre by approx. 40%.
Lighthouse projects within the overall project	Überseequartier as future focus of retail trade and tourism, Elbphilharmonie Concert Hall on the striking Kaispeicher A, International Maritime Museum in the historical Kaispeicher B, Cruise Ship Terminal, HafenCity University.
Reason for the project	Neglected port areas offered the opportunity for city-centre expansion and Elbe River link-up; refinancing of the port expansion in Altenwerder.
General organisational/ institutional conditions	Creation of separate estate Stadt & Hafen, foundation of a city-owned limited company as development company, charged with project development and management of the separate estate.

Source: Dziomba (2009: 115; abridged).

---

4   Only a summary of the project is presented in this chapter. More extensive information with regard to the project creation, planning history and objetives is provided by Penzlien and Bruns-Berentelg (2007), Bodemann (2002) and Schubert (1998). Reference is also made to the HCH homepage (www.hafencity.de), which contains an archive with the most important planning documents in addition to current information on the overall projects, the neighbourhoods, the sub-projects and events.

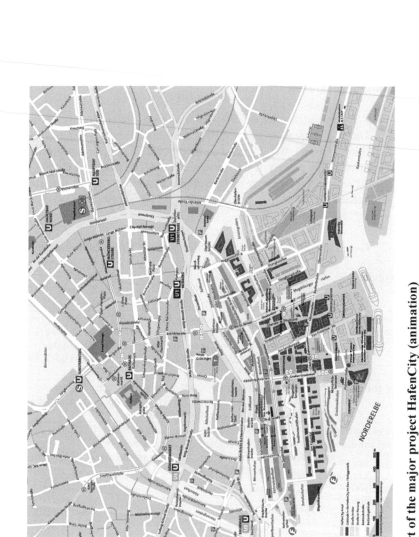

Figure 8.1   Main part of the major project HafenCity (animation)
Source: © HafenCity Hamburg GmbH/lab3 mediendesign.

The objectives that HafenCity pursues are typical of a major urban renaissance project,[5] whereby the high standards demanded of the emerging urbanity and liveliness are emphasised at all times. Above all, there is a wish to restore a residential function to the city centre. As is typical of an urban renaissance project, the urban development structures (compare Figure 8.1) are aligned with the existing city centre and display a distinct variety, structured according to neighbourhoods. In this, a significant emphasis is placed on architecture and suitably designed public spaces (HafenCity Hamburg GmbH 2007b: 4).

HafenCity is located to the south of the city centre, roughly 800 m (as the crow flies) from the Town Hall and 1,100 m from the main railway station; it borders directly on Speicherstadt. A master plan for the major project was agreed upon in 2000, and construction (Kibbelsteg-Brücken) started on this basis in 2001. The first buildings were opened for use in 2004.

The plan is to achieve mixed uses, providing housing for roughly 12,000 people and office space for around 40,000 jobs. In addition, there are plans for retail and gastronomy areas and also cultural and infrastructure facilities.

Aspects of Urban Governance:
Pressure to Succeed, Closure Trends and Legitimisation

Tremendous pressure to succeed characterises the general circumstances facing major projects: the so-called point of no return (Häußermann and Simons 2000: 66) is close at hand and is being talked up by the political opposition in an attempt to bring the entire project to its knees; just one project section can paralyse or even temporarily halt the entire process. In order to legitimise the massive public investments, proof must be submitted continually that the decisions, frequently made in private negotiations with representatives of the private sector, were sensible and correct: there is scant leeway for readjustments if the success of the project is not promptly manifest. This lays bare the 'paradox at the heart of all forays into urban governance' (Einig et al. 2005: VIII): the local authority is reliant on bringing in the private sector and opens itself to cooperation, within the framework of which, however, participation and transparency must be severely curtailed in relations with the general public.

---

5   Jürgen Bruns-Berentelg, CEO at HafenCity Hamburg GmbH, explains: 'If there is a central criterion for success in developing HafenCity, it lies in achieving a high quality of urbanity. It doesn't matter that it is extremely difficult to pin down the term, as we know what it means; our first major step in completing and tangibly sensing the urban quality of HafenCity will be achieved in Dalmannkai, with its high residential quota of 630 apartments, office buildings of different sizes, street-level uses to meet public demand, the framework of broad, richly varied promenades and squares, construction of the all-day school and the Traditionsschiffhafen, which also incorporates the Sandtorkai development' (HafenCity Hamburg GmbH 2007a: 1).

This situation also applies to property sales within the framework of developing HafenCity. Although the various representatives and bodies involved are legitimised by election – e.g. the Commission for Land Reallocation, manned by representatives from the Bürgerschaft, the District Assembly and the city government,[6] rules on the property sales prepared by HafenCity Hamburg GmbH – the sessions are held in private and the contractual provisions kept secret. This enables all those involved to work together in the necessary professional and confidential manner; but the general public, informed of decisions only at a later date, is often unable to trace the logic behind the decisions.

HafenCity counteracts this phenomenon of throughput legitimisation (Haus and Heinelt 2005: 15) by drawing on extensive information and marketing measures, e.g. the information centre Kesselhaus, a wide variety of information leaflets and guided tours of the project zone. In this way, explanations of the general circumstances and the basis for decisions as well as the modes of action are instrumentalised in order to make the rulings comprehensible.

Public Sector Property Sales

As a major urban renaissance project, one of the special features of HafenCity is that the public sector is taking the initiative on areas it owns. This leads to significant differences in the objectives and governance structures in the major project, compared with the development of large-scale, former railway or industrial areas in private sector ownership, for instance: the public sector is indeed the primary entity of planning and project management and is required to uphold the public good and not pursue profit maximisation. It must therefore place high standards on the quality of urban development and urban policies, on the basis of which sustainable intrinsic value and urbanity should be created.[7]

Irrespective of the actual objectives and priorities pursued in the major project, the public sector is bound by specific legal obligations in selling its properties, unlike private sector companies that are entitled to enter into direct negotiations with partners they can select freely. First, in accordance with the respective financial regulations, entities under public law are required to sell the properties in their possession at their full value, as determined by the market value, i.e. the price achievable as per the valuation date. Bills of sale that violate the ban on divestiture below market value may indeed prove to be null and void.[8]

---

6   Compare Land Reallocation Commission Act dated 29 April 1997, available at: www.landesrecht.hamburg.de [accessed: 30 December 2011].

7   It is important to note that HafenCity was initiated originally with the objective of using land development to raise congruent funds for financing a new port location (Bodemann 2002).

8   Compare Usinger (2002: 508); however, exceptions can be reasoned on the basis of municipal tasks and public interest, e.g. the promotion of trade and industry.

There are three fundamentally different ways of determining the market value (Falk 2004: 876): mark-ups on the project developer's economic margins (residual value analysis), alignment with the realty and property market by independent surveyors or estimator committees and also the determination of market value through a bidding procedure. The two first cases require professional property sector expertise on the seller's part – in the first case to apply the correct parameters to the residual value analysis, and in the second case to consult the suitable comparative locations. In both cases additionally, either the foreseeable improvement in market or location circumstances or a project developer's risk premium in view of prevailing uncertainty in the location's outlook must be taken into account. These calculations are *crucial* in determining the success of a project section; after all, they have an overriding influence on the commercial leeway for action – and therefore on the project developer's willingness to comply with the seller's notions concerning urbanity and sustainability. Conversely, the third case, the bidding procedure, comes with the risk that an excessive offer will win as a result of a 'bidding frenzy' or an erroneous assessment of location and market potential. The additional income for the municipal coffers, which may at first glance appear welcome, does, however, heighten the risk that the project section in question will fail (Dziomba 2006: 80). The project developer's limited flexibility forces him into one-sided structuring to achieve the highest-profit exploitation and to target the most solvent groups, which then hamper the conditions for urbanity.

Second, public procurement laws dictate that property sales connected with a construction contract by the public sector must be on the basis of a tender procedure.[9] Although tender procedures were not prescribed in Germany – nor in HafenCity– until the beginning of 2007 (Wagner 2007), the European Court of Justice ruling (EuGH ruling dated 18 January 2007) ('City of Roanne') and the Higher Regional Court ruling (OLG Dusseldorf dated 13 June 2007) ('Ahlhorn Airport') fundamentally changed this perception: a construction contract is always assumed to exist whenever public owners are not exclusively selling property, but are also linking it to construction agreements or urban development contracts. Accordingly, a formal, European tender invitation must take place before any sale if the threshold value is actually exceeded.

Compared with a pre-2007 sales procedure that only satisfied the requirements under state aid laws, action taken in accordance with public procurement laws comes at a higher cost in the sense that the sale objective and the criteria for accepting a bid must be defined in the tender specifications in detail and in a verifiable form. A broad range of criteria is conceivable here to enable a demand

---

9 'Construction contracts must be tendered according to VOB/A both below and above the EC thresholds. On all accounts, if contracting exceeds the threshold, property developer agreements, leasing agreements or construction permit agreements count as construction contracts and not as other services that must be tendered according to VOL – provided that in them, construction services are provided for a public sector principal' (Jasper and Marx 2007: XXVI).

for urbanity, to enshrine the mutual objectives of major projects and to define clear instructions with regard to the desired concept variety, mixed use and sustainability standards.

Steering towards Urbanity

Before considering examples of two governance instruments employed in HafenCity in order to 'force' project developers into laying foundations for urbanity, it is important to clarify which measures these are precisely.

Laying Foundations for Urbanity

One of the requirements for urbanity in mixed uses is that the areas are indeed suitable for different purposes; for instance, retail trade and gastronomy require higher ceilings, additional exits and ventilation, etc. Varying sizes are favoured, depending on the use, concept and operator. A certain level of flexibility is advantageous here. Office and surgery space in particular must be suitable for alternative uses, i.e. they must not be tailored too precisely to suit specific users and their needs.

Social mix and a high degree of liveliness are achieved on the one hand by a large, multifaceted range of housing in all price categories, sizes and ownership forms. Alternatives to the top-price segment become possible if developers realise that this concept of variety, and not their bidding price, is decisive. On the other hand, retail trade, gastronomy and cultural institutions, just like 'crowd pleasers' such as museums, concert halls and other events, must address a variety of target groups.

Diversified architecture need not come with tiny plots that project developers dislike; sizes designed with the needs of the people in mind can also be created in larger units by instrumentalising 'internal parcelling'. There will be greater courage to seek diverse creative solutions if the targets are transparent.

Governance Instrument 'Market Segmentation'

In keeping with urban renaissance, HafenCity Hamburg comes with the aspiration of creating propositions for a broad spectrum of users. These quality standards placed in urban development, architecture and the design of open spaces were satisfied in the case of the Dalmannkai/Kaiserkai, as were expectations with regard to a heterogeneous and small-scale mixture of uses and users, by defining detailed instructions in the tender specifications. The plan was to force project developers interested in short-term profits to come up with elaborate and, if necessary, unusual concepts of combining several uses under one roof and to see themselves as building blocks in an 'overall composition'.

In order to achieve this, the development area was divided into several properties, simultaneously offered for tender, thus providing project developers with sensible, i.e. efficient, surface scales. The development orders divided the properties into smaller architectural and conceptual units and permitted small-scale real partition for subsequent sales. A diverse mixture of market segments and ownership forms was designated in order to prevent a situation within which all development orders would have a similar concept and consequently generate competition for the same user, operator and buyer groups: classic apartment complexes alongside luxury designer condominiums, freehold apartments in different sizes and price categories, housing for senior citizens and also families; housing cooperatives and groups were also included. The ground floors house restaurants, cafés, shops, galleries and offices.

This mixture was made possible as the tender specifications did not define any fixed requirements for capacity utilisation and use quotas for the properties – quite the contrary, the prescribed, staggered prices per form of use, the wide range of desired uses and the clear definition of selection criteria indicated to the bidders their leeway in developing an individual concept. Ultimately, by selecting the bids, the tendering party 'composed' a real neighbourhood with complementary projects.[10]

Governance Instrument 'Market Calibration'

This instrument was developed for the 'prime locations' within HafenCity and was implemented as an 'artificial cut on supply'. Large properties that would be suitable for prestigious corporate headquarters, but are scarce in the small-scale Hamburg city centre, are rarely tendered for sale, thus enabling price and concept-maximising competition among potential promoters of local industry within a select group.[11] Prospective buyers gain an edge on their competitors in negotiations by satisfying the demands for urbanity and sustainability.

This procedure is advantageous in that the major project HafenCity not only prevents a feared saturation of the office property market in Hamburg, it actually reverses it: the heightened demand for large properties among companies wishing to move within or to Hamburg due to the positive development in the urban renaissance project is satisfied by the supply of just a few suitable properties, so that short supply can lead to higher prices and greater concessions in terms of the

---

10   In the Dalmannkai example, HafenCity Hamburg GmbH retained the right in the tender specifications to transfer a bid submitted for a certain plot to another plot and to offer development of this alternative plot to the bidder, if it proved suitable for the overall concept. This example demonstrates the instrument's particular flexibility.

11   The so-called case of promoting local trade and industry should be named as the special case, prevalent in Hamburg, that provides exemption from the obligation to invite for tender and permits free contracting. This requires that at least 50% of the office or commercial space will be occupied by a user named in advance for at least 10 years.

seller's objectives. However, in addition to high planning security, this procedure presupposes that a recognised location quality has already emerged in the environs of the properties in question.

Summary

Urbanity does not emerge out of thin air, especially in the case of major, high-price, city centre projects that are subject to a high pressure to succeed. Nonetheless, HafenCity Hamburg shows how at least some requirements for the creation of urbanity can be secured in the phase of property sales by drawing on a suitable contracting procedure and adequate individual process design. At the same time, this contributes to the success of the project sections and therefore also to the success of the project as a whole, also reducing the risk to the buyer and the seller, i.e. to the public sector.

The strategies must be aligned with both the general circumstances of the overall project itself and also with the requirements of each project section and the local property market. When facing a strong macro-location with the corresponding size and variety of demand on the property sector, instruments such as market segmentation and market calibration can, given clear specification of the contracting criteria, contribute to inducing interested project developers to compete to a greater extent through the fulfilment of desired urbanity and sustainability aspects, and not just in the bid price they offer.

However, it is important to note, with regard to the potential to transfer this concept to other cities, that Hamburg most certainly offers an extraordinary general set-up. Given the extraordinary quality of the macro- and micro-location and the market as a whole, an exceptionally professional quango team with strong negotiation skills can operate flexibly on the market with the backing of the political and administrative sectors here. But every major project stands and falls, at least in part, with its 'lighthouses'. Accordingly, at least since the onset of the economic and financial crisis in 2008, the 'flagships' Elbphilharmonie (Kähler 2008) and Überseequartier have shown clearly that the selection and contractual integration of project partners are of immense significance in these major, more prominent and complex project sections especially, given that different market phases will have to be endured over the course of their longer development periods.

There is a need for further research – especially with regard to the urban planning practice and the ownership circumstances concerning wasteland – on ways in which the public sector could exert influence on projects involving privately-owned property. Profit-seeking will play a larger role in the revitalisation of this kind of former rail, post, telecom, industrial and commercial land than urban development or social aspects, even if the property sector does, of course, look favourably on the emergence of urbanity as a future location factor.

References

Bodemann, U. 2002. HafenCity Hamburg – Anlass, Masterplan, Chancen, in *Hafen- und Uferzonen im Wandel: Analysen und Planungen zur Revitalisierung der Waterfront in Hafenstädten*. Berlin: Leue, 99–117.

Bodenschatz, H. 2005. Vorbild England: Urbane Renaissance in Birmingham und Manchester. [Online] 3. Available at: www.kunsttexte.de/index.php?id=711&i dartikel=12360&ausgabe=12137&zu=121&L=0 [accessed: 12 January 2011].

Dobberstein, M. 2000. Das prozyklische Verhalten der Büromarktakteure. Interessen, Zwänge und mögliche Alternativen. *Arbeitspapiere zur Gewerbeplanung* (2), Dortmund.

Dziomba, M. 2006. Großprojekte auf innerstädtischen Brachflächen. Revitalisierungs- und Vermarktungsprozesse und ihr Einfluss auf den Projekterfolg. *Berichte zur deutschen Landeskunde* 80(1): 65–84.

Dziomba, M. 2009. *Großprojekte der urbanen Renaissance. Die Phase der Grundstücksverkäufe und ihr Einfluss auf den Projekterfolg*. Münster: Lit-Verlag.

Einig, K., Grabher, G., Ibert, O. and Strubelt, W. 2005. Urban Governance – Einführung. *Informationen zur Raumentwicklung*, 9/10: I–IX.

Falk, B. 2004. *Fachlexikon Immobilienwirtschaft*, 3. Auflage. Köln: Immobilieninformationsverlag Rudolf Müller.

Glaser, B.G. and Strauss, A.L. 2005. *Grounded Theory. Strategien qualitativer Forschung*. Bern: Huber.

HafenCity Hamburg GmbH (ed.). 2007a. *HafenCity News*, 9, June. Hamburg.

HafenCity Hamburg GmbH (ed.). 2007b. *Projekte – Einblicke in die aktuellen Entwicklungen*. 7. Auflage, March. Hamburg.

Häußermann, H. and Simons, K. 2000. Die Politik der großen Projekte – eine Politik der großen Risiken? *Archiv für Kommunalwissenschaften* 1: 56–71.

Haus, M. and Heinelt, H. 2005. How to Achieve Governability at the Local Level? In *Urban Governance and Democracy*, edited by M. Haus and H. Heinelt. London and New York: Routledge, 12–39.

Jasper, U. and Marx, F. 2007. Einführung, in *Vergaberecht*. München: Beck.

Kähler, G. 2008. Unglückliche Partnerschaft. Erst wollte Hamburg den Konzertpalast aus Spenden finanzieren. Jetzt explodieren die Kosten. *Die Zeit*, 25 September.

Penzlien, B. and Bruns-Berentelg, J. 2007. HafenCity Hamburg – Qualitative und quantitative Dimensionen des Großprojekts der Innenstadtentwicklung, in *Metropole Hamburg – Projekte zum Leitbild 'Wachsende Stadt'*, edited by C. Krajewski and R. Lindemann. Münster: Arbeitsberichte der Arbeitsgemeinschaft Angewandte Geographie Münster e.V., 36, 29–58.

Schubert, D. 1998. HafenCity Hamburg. *RaumPlanung* 83: 211–22.

Schubert, D. and Klein, M. 2006. Quago/Quango, in *Das Politiklexikon* [Online]. Available at: www.bpb.de/popup/popup_lemmata.html?guid=WGH363 [accessed: 12 January 2011].

Urban Task Force 1999. *Towards an Urban Renaissance. Final Report of the Urban Task Force.* Chaired by Lord Rogers of Riverside. London.

Urban Task Force 2005. *Towards a Strong Urban Renaissance. An Independent Report for Members of the Urban Task Force.* Chaired by Lord Rogers of Riverside.

Usinger, W. 2002. Rechtliche Probleme bei Kauf- und Gewerbemietverträgen für Entwicklungsobjekte, in *Handbuch Immobilien-Projektentwicklung*, edited by K.-W. Schulte and S. Bone-Winkel. Köln: Rudolf Müller Verlag, 489–532.

Wagner, O. 2007. EuGH: Ausschreibungspflicht bei städtebaulichen Entwicklungsvorhaben. *Immobilien-Zeitung*, 8 March.

Chapter 9

Neighbourliness in the City Centre: Reality and Potential in the Case of the Hamburg HafenCity

Ingrid Breckner and Marcus Menzl

Introduction

The development of city centres has increasingly been in the public eye in recent years. While office and retail buildings have characterised the face of many cities for several decades, living space has since gained ground as a desirable usage. The idea of living in the city brings with it complex expectations. People hope that the city will be established not only as a place of business but also for their private everyday lives. The urbanity and diverse public life which are inseparable from the inner city are, according to such ambitions, not to be achieved by staging events nor to close down when the shops close, but should result from the complex and possibly conflict-ridden layering of different usages, diverse interests and contrasting patterns of behaviour.

Introducing such a fundamental transformation is, however, no simple task. There is no doubt that the renewal of city centres can only be achieved by means of establishing more living space if the new residents build up local attachments and neighbourly structures. Precisely because of the substantial presence of other usages, living in the city centre in a private and individually focussed way, without real neighbourhood life, would not have any great effect on the character of the city centre: there would not be higher demand for local retail and restaurants, no processes of laying claim to public space and no new well-developed local representatives.

The HafenCity can serve as an empirical example for the implementation of a complexly conceived city centre, because there is decisive development of dense urban use of space by visitors as well as spaces characterised by everyday life and neighbourly accommodation. Concretely, this can be seen in the fact that, beyond 45,000 jobs and the development of a large number of reasons to visit the HafenCity (the Philharmonic Orchestra, the cruise ship dock, several museums, many opportunities for shopping and dining as well as short-term events), 5,800 residences have been built along with the corresponding infrastructure, services, recreation areas and opportunities for neighbourhood networking (please see www.HafenCity.com for further details on the project). In other words, the

HafenCity seeks to realise both urbanity and neighbourliness in a symbiotic relationship or at least a sort of co-existence.

This task is not only highly attractive, but also very challenging, and it brings forward several questions. How can the one-sided dominance of classic city-centre usages be relativised and long-term neighbourhood life in the city centre be established? Are the aims of urbanity and neighbourliness compatible, or do they involve fundamental conflicts at several points, such that any solution must compromise one or the other? Can these processes be guided along, and if so, how? Along with the question as to whether urbanity is planable comes the question as to whether neighbourliness can be planned, or if it is something that must develop along its own lines. And finally, there are several studies which portray neighbourliness as an outdated concept. The door-to-door communities which were a matter of course for such a long period, and into the 20th century, and which involved close contact with those people living in the immediate vicinity, have, according to these studies, lost in significance because of increased mobility and individualisation. This is particularly the case in large cities. These forms of community were replaced or augmented by other models. Most of all, one can observe an increased permeability of space such that socially significant locations for particular individuals are at a considerable distance from each other as a result of the development of transportation and the development and usage of new communications media. This leads to the development of place-to-place communities. Recently, it has even been possible to observe the increased relativisation of attachments to specific locations, since people could contact each other using mobile phones or the Internet regardless of where they were (a person-to-person community) (Straus and Höfer 2005). Does this make the neighbourhood passé? Or to put it in more general terms: how do social relationships, of whatever kind, work within a local context and how can they be conceived of for inner-city urban contexts?

The argument of this chapter proceeds in four steps. First, we will discuss the thesis that social processes are always pre-structured by planning, whether explicitly or not, on the basis of empirical findings from the HafenCity Hamburg. This applies both to urbanity as well as to forms of networking, the formation of neighbourly relationships, or the development of social identification with the area. Following on from there, we will present the full range of neighbourhood life in the new HafenCity. What motivates which local actors in the HafenCity to get involved in their neighbourhood in which ways? How were neighbourhood activities initiated and which conditions facilitate these processes? In a third section, we will analyse the newly established neighbourhoods typologically. Does the HafenCity include novel patterns of urban neighbourliness? What are their particulars and how can they be explained? In closing, we will summarise the most central findings according to our initial question and discuss the issue of how far the neighbourhood structures developing in the HafenCity, as part of the city centre, can be compared with those of other quarters.

The Pre-structuring of Social Processes through Planning

In this chapter we are proceeding on the basis of findings indicating that social processes are pre-structured in manifold ways by the specific character of the space in which they take place. This follows Martina Löw (2001: 158–9), who distinguishes between two different processes of constituting space. These are the act of spacing, which includes the construction, building or positioning of goods, symbols or people, and the achievement of synthesis, that is, the perceptive, imaginative and memory processes of people, who parse the various goods and people into spaces. Both processes stand in a close reciprocal relation to each other and are equally responsible for the constitution of spaces. This is important because spaces can no longer be understood, as was long the case, to be absolute orders primarily arising out of planning. Conversely, this also means that the significance of the acts of the people using the space cannot be absolutised either. Planning decisions colour spaces and constitute a marked component of the spacing process which pre-structures social behaviour.

In the following we therefore defend the thesis that the consequence drawn from the appropriate emphasising of individual processes of perception and identification should by no means be a reduction of pre-structuring planning (as part of the spacing process), but rather, an ambitiously refined, far-reaching and complex planning concept must be articulated. It is not, on this view, the task of planning to produce finished spaces nor to attempt to determine social behaviour patterns, but instead to provide stimulants for intensive and independent social processes. This need not necessarily happen by means of less planning, but, among other things, by means of more planning on the part of different actors in the various relevant fields of action and with specific aims which, in an ideal situation, can be adapted to changed uses. In the following this type of pre-structuring social processes through planning will be illustrated on the basis of the example of the HafenCity.

Creating Occasions, Mixing Functions:
The Production of an Atmosphere of Interest

The HafenCity was conceived as a functionally complex, urban city centre with an atmosphere of interest (Böhme 2006: 105), which distinguishes itself from monofunctional structures of pure office blocks or residential areas. The implementation of extremely small-scale functional mixing, which in many cases reaches inside the buildings, is connected to this. Most of the ground floors are reserved for public usages (restaurants, retail), and in some cases buildings have office or residential usages distributed across different floors or in vertical chains. The goal of mixing is to create as many occasions as possible to come into the HafenCity and spend time there. A casual observer cannot immediately tell if someone is there on the way to work, shopping, or to make a visit, or for some other reason. He encounters both familiar and strange faces, accustomed and surprising behaviour patterns. The conditions for anonymity in the sense of freedom to act as one pleases and the development of individuality are thus established (Simmel 1903/1995).

The variety of occasions to engage the HafenCity results in not only a high degree of diversity among users and usage aims, but also allows the quarter to remain busy, since the desertedness of low-usage times is relativised (Feldtkeller 1994: 57). The density and shifting foci of different patterns of behaviour throughout the day enrich the quarter and create an impetus to go out onto the street and spend time out. What is decisive for the interesting atmosphere in the HafenCity, and also a result of planning decisions, is the intensive architectural referencing of the water and the sharp contrast between the high building density and the open water. There was a conscious decision not to raise the promenades onto the level of the high tides in order to allow people to walk along the Elb even at low tide. In addition, ways were found to accent the maritime character of the HafenCity and allow people to experience it intensively, such as the Pontons in the Sandtorhafen, the traditional shipping harbour, and the integration of the cruise ship terminal. This shows how apparently purely engineering questions such as flood protection can have far-reaching implications for the perception and usage of spaces (for more details see Bruns-Berentelg 2010).

Along with the small-scale layering of usage aims and interests, there is also the presence of friction and conflict. A certain amount of confrontation resulting from the juxtaposition of flats and restaurants or a basketball court have been deliberately taken into account. They are considered an elementary component of an urban environment and constitute a further occasion to relate to the public sphere, whether as observer or as critic.

Deliberate Choice of Property Owners:
A Broad Mix of Life Situations, Milieux and Lifestyles among the Residents

In the HafenCity the composition of the body of residents has been influenced by diversification of property owners and the residential market. This can be seen in the presence of different social groups, although they are mostly within the upper and middle classes. In future, when subsidised housing has been completed, the range of residents can be expanded. Building property was not granted to the highest bidder, but rather on the basis of different set prices, and that has been of central significance for achieving residential diversity. Investors were chosen on the basis of the conceptual quality of their proposal. Concretely, that means asking how the investors intended to meet the requirements stated in the call for proposals, such as family-friendliness, usage of the ground floor, sustainability, letting and purchasing prices, etc. Which target groups could be approached with their proposal, how would their proposal fit in with the whole quarter? With the help of this procedure, it was possible to include a broad set of different property developers and different sorts of flats in the HafenCity. Besides property developers who built and sold condominiums, private individuals were also able to invest in several cases and build flats and maintain them among their investments. Further, building societies also constructed flats, cooperative projects were realised, and flats were made available for specific target groups such as pensioners, students or musicians.

The body of residents in the HafenCity reflects the diverse supply of flats and is by no means limited to wealthy households, but also includes a substantial proportion of average and middle-class households. Often these are two-income households, which do not yet include children or whose children have already started their own household. The proportion of households with children is now already at 12%, a level which is comparable to the popular central neighbourhoods like Neustadt, Winterhude (both 11%), Elmsbüttel (12%) or Hoheluft-West and Ost (each 13%) (Statistisches Amt für Hamburg und Schleswig-Holstein 2010). This has been facilitated by the displacement of the primary school from the old city into the HafenCity along with conceptual reforms, the creation of attractive child-care facilities, green spaces and safe outdoor areas in the direct vicinity of the flats. This diversity was also deliberately encouraged in that calls for proposals required developers to accommodate a high proportion of family-friendly flats. In general, these strategies served as the means of influence necessary to ensure the presence of all age groups and a reasonable degree of diversity in lifestyles and life situations.

The mix of types of developers is also important in order to ensure (on the basis of the sales contract) that a substantial proportion of owners either keep the flats built (as in the case of building societies) or use them themselves (as do building cooperatives) and refrain from selling them for profit. By this means a certain degree of continuity and a consistent interest in the positive development of the quarter can be achieved. Experience shows that particularly those residents living in flats belonging to a building society or a cooperative are especially engaged in developing neighbourhood relationships in the quarter – another reason to take special account of this type of investor.

Differentiating the Private and Public Spheres

The goal of developing a dense urban city centre requires one to avoid certain types of buildings. There will be no family homes with private front gardens or villas, but only multi-storey blocks of flats. Open spaces are thus mostly of a public or semi-public nature, as in the case of inner courtyards which are reserved for residents of the adjacent buildings. Private space ends at the balcony or terrace. All plazas, promenades and footpaths are of a public nature or, even if they are private property, are equipped with public access rights. The one-storey office buildings, like that of Unilever, are also publicly accessible. This ensures that the HafenCity has a well-developed public character and that in principle the entire public area is open to both residents and visitors.

By means of measures like festivals which serve the process of place making throughout the region and beyond, but also by integrating a basketball court and deliberately facilitating skateboarding in public squares, attempts have been made to provide for those social groups who are not able or willing to live in the HafenCity or who do not identify with the sophisticated design of the HafenCity. These examples also show the importance of a strong planning and regulating hand in areas with diverse users and, at times, starkly contrasting interests. Without such protective intervention there would be a danger of individual residents or business owners with a large investment capacity achieving their own interests and chasing away those less powerful usages like basketball or skateboarding which are important for the character of the area, because they perceive them as involving the presence of outsiders making a lot of noise (Menzl 2010). The need for open spaces, which are available 'for spontaneously and freely-chosen activities which are not pre-determined by any other prescriptions or descriptions' (Spiegel 2003: 176) does not necessarily ensure free access to all interested parties. It requires the integration of planning and regulating intervention in order to make those usages possible which are important for a lively neighbourhood but which could be resisted by some.

The classic function of public space as the area where people come into contact with each other, interact and express themselves, in which intense social processes take place and, basically, everything happens which is meant by the term urbanity, is being supported in the HafenCity. Even if public space in general is subject to diverse changes and shifts in significance (Selle 2004; Wolfrum 2010) and the HafenCity includes some exceptional elements (due, for example, to the emphasis on public space because of organised events of various magnitudes), public space still constitutes an informal meeting place and a place of communal activities and thus a central sphere for the development of social relationships within the local context. Public space is able to fulfill this function especially well because of its gradation. Besides public spaces, semi-public spaces were also created which are accessible to specific residents, such as inner courtyards, collectively used roof terraces, large stairwells (in which concerts have already taken place) or common areas which were integrated into two residential buildings and which serve as a meeting place for the immediate neighbours.

Pre-structuring these spaces through planning, as can be seen from the examples discussed here, is far from determining individual behaviour, but it provides a point of departure which facilitates both an urban atmosphere and forms of neighbourhood cohesion. The social relationships of residents of the HafenCity will be considered in more detail below.

Social Relationships in the Local Context of the HafenCity

Living in the HafenCity is not usually a compromise solution which people have found when looking for a flat, but rather the result of a deliberate decision to move (Menzl et al. 2011). Accordingly, a large proportion of the residents arrive with a positive and open attitude, but also with high expectations of the HafenCity. From this point of departure, diverse forms of neighbourly interaction developed very quickly, along with a great degree of interest in getting involved in the new neighbourhood. Communication platforms (such as a digital residents' forum, the HafenCity newspaper) and networks quickly arose out of the initiative of residents. These have been growing continuously from the start and they are an integral part of the integration of new residents, especially because of the corresponding activities (twice-monthly neighbourhood meetings, a flea-market, summer festival, various block parties and excursions).

More formal types of neighbourhood involvement have arisen out of the communication networks just described. The spectrum includes everything from the Kunstkompanie HafenCity e.V., a club which implements artistic and cultural projects of various kinds in the HafenCity, to the sport club Störtebeker SV and the club Spielhaus HafenCity e.V., which has taken responsibility for the building and maintenance of a playhouse on the Schatzinsel playground.

In order to further support the co-responsibility of residents and businesses in the HafenCity, the club Netzwerk HafenCity e.V. was founded in autumn 2009 and has the potential to act as a neighbourhood council. Many detailed questions which affect everyday life in the HafenCity are now being discussed and organised by this club within the framework of various task groups. The club sees itself explicitly not as a forum in which particular interests of individuals can be realised, but rather as focussed on resolving issues of collective interest with reference to the philosophy of the new neighbourhood. The founding of the club was thus an important step on the way to greater self-regulatory structures, which, given the high degree of cultural capital in the neighbourhood, is a realistic goal.

In sum, there is an extraordinarily diverse range of local networking activities in the HafenCity, especially relative to the number of residents so far and the central location with its many alternative activities. How can this be explained? There are reasons to believe (and these reasons will be outlined below) that the multifarious social life in the HafenCity is the result of a primarily constructive and generally intensive matching process, coordinating the expectations and wishes of the residents, businesses and social institutions, on the one hand, and the newly developing neighbourhood with its particular philosophy and characteristics, on the other hand.

This is evident even in the motivations for moving to the HafenCity. In many accounts, a desire for a new start is of central significance, whether on the part of empty-nesters, who are re-orienting themselves in reaction to the reduced size of their households, or pensioners, who deliberately want to attempt a new start, or young families or couples, who are buying their first flat. The biographically determined desire for a more or less pronounced break is implemented by moving to a completely new neighbourhood, in which things are in transition and can be influenced and, because of its central location, is still, at least for those from Hamburg, familiar, even if not as a place of residence. Fascination with this doubly new way to live motivated many residents to publish their personal experience of the neighbourhood over the Internet and share it with others. Photographs were put online on various websites, and the HafenCity-News was originally a digital neighbourhood paper, founded, incidentally, by a resident who had no previous journalistic experience but was simply interested in experimenting with a new role. Other residents were inspired by the large stairwells in their buildings and put on concerts. This enthusiasm for a new start was taken up and supported by the development society, the municipal HafenCity Hamburg Ltd. There is an effort to convey the philosophy behind the HafenCity to new residents and to support the realisation of the project by means of welcome packages with important information on the history of the place as well as all relevant information on everyday life in the new neighbourhood, regular information and discussion events and very high-density communication. Identification with the new place of residence, which is already present in many cases, can be strengthened in this way. Local contact is given a great significance, not least because of the high-profile and frequently controversial perception of the HafenCity in the city as a whole. Even if they wanted to, it would hardly be possible for the residents to see their neighbourhood as an ordinary and interchangeable place to live.

The second point to stress is that the development of social life in the HafenCity and the networking of local parties has profited from several spatial and institutional nodal points, which it was possible to create as a start-up help. For example, a building society included a common room in their building in the Sandtorkai/Dalmannkai area. The responsibility for planning the events that would take place in this room was given to a foundation established by the society. After only a few events the residents took over planning of the monthly neighbourhood meetings and further neighbourhood activities. The room has since then hosted a large number of meetings for planning parties, founding clubs and for neighbourhood parties. The Spielhaus HafenCity e.V. club also began as a result of an external stimulus. In 2008, the HafenCity Hamburg Ltd. invited all of the families living in the HafenCity, along with primary school students, to plan a playground. In this connection several parents expressed a desire to have a small playhouse on the playground. This wish was accepted by the HafenCity Hamburg Ltd. on the condition that the parents took responsibility for maintenance of the house and for part of the costs. The parents then founded the Spielhaus HafenCity e.V., collected donations from nearby businesses, and have since guaranteed the use of the house, which in fact was designed by one of the parents, who is an architect. This example shows how:

- networking between households in similar stages in their lives can be supported;
- residents can be encouraged to take an active role in claiming public space (even in a central context with many competing usages); and
- responsibility for the development of the HafenCity can be distributed to multiple parties.

What can be seen from this example is, not least, that the specific way to the formation of social life in the HafenCity was characterised by the high social, cultural and, in some cases, also economical capital of many residents. Such resources allowed for clubs to be formed and common activities to be relatively easily organised (such as neighbourhood parties for all residents and those who work in the HafenCity at the cruise ship docks). Independently of these concrete advantages, neighbourhoods like the HafenCity also profit fundamentally from the fact that people with a high level of education and secure incomes are more likely to volunteer and get involved than other segments of the population (Hoch and Otto 2005: 498; Trunec 2010: 15).

Finally, a third point should also be mentioned. Common everyday experience is a very significant aspect supporting local networking among the residents. It is no coincidence that, in the first months, people frequently described themselves as pioneers. Many things were unfinished and had to be improvised, and this situation was shared with others. Disgruntlement about the four-lane design of the street Am Sandtorkai was a similar sort of experience and was first articulated at a welcome event hosted by the HafenCity Hamburg Ltd. Conversations between neighbours about the HafenCity resulted from this, as well as the formation of the Internet forum hafencityleben.de. This communication medium led to direct contact with the relevant planners so that several small modifications of the plans could be achieved and with them improvements in the everyday situation of the residents. On the other hand, and this is even more important for this argument, intensive communication networks are a deciding precondition for local and neighbourhood involvement, since they provide contact points for individual activities and strengthen people's confidence that the investment of their time will indeed have an effect. They thus also play a large role in the self-reinforcing effect which is characteristic of functioning and active social networks.

In closing this section, it should be mentioned that there are also residents in the HafenCity who, whether due to disinterest or other priorities, remain withdrawn from the local context and tend to avoid the community and the urbanity of public spaces. Such valuation of privacy and anonymity are completely legitimate expectations for life in the city, since the city, unlike the traditional village, allows the individual to choose between openness and encounter or indifference and withdrawal. Despite all the significance that urban encounters and confrontation with strangers and the production of a public sphere has for the development of the city, one must not forget that the opportunity for encounter must not be allowed to become a requirement of encounter in the sense of undermining privacy as in a traditional village (Schroer 2006: 245).

Disinterest in the local context only becomes problematic when personal wishes regarding quiet, views and exclusivity are imposed on public space and far-reaching demands for the reduction of urban complexity are made. The city is then no longer being understood as a place in which diverse patterns of behaviour and claims to usage coalesce, but rather as an imposition which restricts personal freedom. In general, these individuals do not engage their neighbours in dialogue or look for compromises, but push through their interests by means of high-level interventions with decision-makers, public and media pressure, or even in the courts. Such anti-urban fixations on personal interests and NIMBY attitudes can be found in many places in the city and are by no means a phenomenon particular to new inner-city neighbourhoods. This attitude can also be found in the HafenCity, but has not (yet) achieved a majority, since most of the residents identify with the HafenCity project which includes shared responsibility and consensual solutions for problems which still need to be resolved or conflicts which may arise in future (Menzl 2010).

Neighbourliness? Networks? Which Social Frameworks are Forming?

The discussion above has made clear that social life in the HafenCity is characterised by a high degree of liveliness, willingness to get involved, and very intensive forms of local identification. This is initially surprising, not only for outside observers. Individual residents also say that the intensive forms of neighbourliness that have developed in the Hafencity were astonishing and completely unexpected. Some of the most involved persons emphasise the fact that they had never previously got involved in their neighbourhood, but are now enjoying doing so. The question which remains open is what exactly is developing in the HafenCity: neighbourliness, networks, or something else entirely?

As described initially, neighbourhoods are often described as an outmoded concept, which no longer fits in with present social realities. The classic neighbourhood involves intensive face-to-face contact, a broad system of reciprocal help, and for a large body of knowledge of each party about the others. This is stronger in a small village than in a city, which tends towards anonymous structures. Not only the changed communication patterns mentioned above (Internet and mobile phones), but also, more generally, the massive processes of individualisation in post-Fordist societies lead to changed forms of social life. Three factors are particularly noteworthy in this regard (Straus and Höfer 2005). Firstly, social attachments which are specialised and limited and, unlike in the traditional neighbourhood, do not reveal the whole personality, are preferred. Secondly, dense, close communities are increasingly being replaced by loose, often very heterogeneous networks composed of low-key relationships which are characterised by a high turnover of their members. This bears witness to the fact that the majority of people today are involved in multiple complex social spheres so that focussing on classic stable networks would not offer sufficient support. Thirdly, many people tend to engage in community life not in public places, but at home or from home and over long distances (using the Internet or the telephone). As a result, so the argument goes, social life has separated itself from territorial structures and while social activities take place in specific places, they are part of social worlds which extend across local and national borders (Straus and Höfer 2005: 486).

If one looks at the social relationships developing in the HafenCity as described above, the picture that results does not correspond with the assessment of network researchers. To begin with, a substantial portion of the residents are oriented towards traditional patterns of neighbourliness. Especially the residents of buildings constructed by the building society value adherence to rules, helpfulness, direct exchange and a certain degree of commitment, all attributes which are generally ascribed to village or small-town contexts. Even the institution of having a little gossip with the neighbours, insistence on familiar norms, and forms of exclusion of outsiders among the residents exist in the HafenCity especially, but not only, among members of the building societies (on these patterns of neighbourliness see Menzl 2007). Certain classic social needs are re-constituted within the framework of the HafenCity and give the residents a sense of well-being and security and allow for a high degree of familiarity even in the midst of urban distance.

Secondly, the example of the HafenCity shows that the details of the purported trend in developments are far more complex and contradictory than they first appear. While the tendency towards digital networks is also characteristic of the HafenCity, it is augmented by a valuation of the expansion and maintenance of face-to-face contact and concrete local attachments. The networks that have arisen so far do not, at least among the core members, show a high degree of fluctuation, but are characterised by continuity and a sense of responsibility. It is also interesting to consider the motivation for forming a network. In network research, one often differentiates between goal-oriented and non-goal-oriented networks. Networks with a central theme and defined tasks and goals are considered goal-oriented networks. Non-goal-oriented networks are those which have a more general role in the everyday life of the individual and support it at times, but independently of specific projects. A goal can, in these cases, be the formation of 'coherent communities' (Diller 2002: 58). In the HafenCity, both types of networks can be observed. There are coalitions formed to reach specific goals (such as the Spielhaus HafenCity e.V.), but also those which are intended for networking itself (such as neighbourhood get-togethers). It is striking that the combination and layering of both types of networks is the rule. Possibly this is an essential reason for the networks functioning as well as they do. The relative strengths of each type of network can be realised. In the case of the goal-oriented networks, this is mainly the attachment to concrete tasks and simple opportunities to get involved depending on each personality, while the non-goal-oriented networks are characterised by intensive personal commitment and a high degree of trust.

And finally, and this is certainly the most weighty point, the trend in the HafenCity towards networks is not connected to a decrease in the significance of the territory, but rather the opposite is the case. The local context is not becoming meaningless, but is the element which binds the residents together and fundamentally contributes to their connection in networks. Thus, in the HafenCity, we are seeing extraordinarily strongly locally attached social networks. Although the residents generally have numerous alternatives, it is apparent that, for a large number of the residents, there is a desire to get involved in their local area. In other words, there is a wish to share the experience of living there, to take responsibility and in doing so to achieve a sense of belonging and ultimately to develop a 'communal cultural identity' (Castells 2002: 66). It should be emphasised in this connection that the active residents and especially the members of goal-oriented networks have so far consistently focussed on the implementation of collective interests, which are in line with the development of the neighbourhood. The joint implementation of individual interests has, as yet, proved to be an inviable occasion for the formation of networks.

In general, the findings suggest that neighbourliness, understood as social cohesion with a strong local attachment, is not dissolving but is taking on a new appearance. Neighbourliness as it is developing in the HafenCity is based on local networks and is certainly more non-committal, specialised, flexible and open than the traditional neighbourhood network. Neighbourliness in the HafenCity is less focussed on personal advantages and aid in dealing with everyday challenges and only secondarily oriented around the principal of reciprocity. Instead, it is mostly dependent on the collective interest in making the HafenCity an attractive place to live and is manifested in various activities which serve, directly or indirectly, to achieve that goal. Seen in this context, the present state of the HafenCity neighbourhood is one of pronounced networks. This type of neighbourliness, which is presumably more pronounced in newly developed areas, should not be seen as an inflexible framework. Like other forms of neighbourliness, it is subject to a constant dynamic of change and must adapt to new challenges. Whether that can be achieved more successfully than in traditional models of neighbourliness requires its own scientific study as well as everyday observation.

Conclusion: Neighbourhood Life between Planning Interventions and the Development of Independent Dynamics

There is generally a marked scepticism in the literature as to the possibility of establishing a lively and actively self-organised social life in the inner city, whether the topic is neighbourliness or any other sort of network. Dangschat, who elsewhere explicitly notes the development of 'new neighbourhoods' in certain social classes and ways of living (Dangschat 2010: 7), expresses grave doubts in relation to the development of neighbourliness in inner-city contexts:

> In these (inner-city) quarters live people who are not particularly interested in the community life of their building or on their street and who are not very attached to the local economy and whose networks extend far into the city or even into other cities. In the context of such a reserved mentality, one sooner finds tendencies towards withdrawal and isolation; these normally lead to multifunctional but socially homogenous strongholds for some of those in the modern service industries. (Dangschat 2009: 256)

Other authors refer to the difficulty, especially in newly developed areas, of establishing neighbourliness. Empirical findings from London by Mark Davidson (2010) show how neighbourhood relationships are dependent on the time one lives in the area: unlike those in older buildings, people who lived in newly developed parts of London manifested an attitude towards neighbourliness which ranged from distant to hostile and identified more with the property they had bought and its location than with potential neighbours at whatever distance. How can the developing neighbourliness in the HafenCity be oriented to such findings? Are, as Dangschat suggests, the 'wrong' people for neighbourhood bonds living here?

Of course, there are also those residents in the HafenCity who are pursuing ways to distinguish themselves and isolate themselves from local life. The real complexity of the social structures among the residents, however, mitigates against drawing the conclusion that there is a general disinterest in local social relationships. As shown above, networks which are closely attached to the location are developing, and these allow for a specific type of neighbourhood based on given preconditions which correspond to the needs and wishes of many of the residents who have chosen to live there. Two points should be especially emphasised in closing.

Unlike many other urban quarters, the HafenCity generates a surplus of significance which goes far beyond the fact of residence itself. The HafenCity has a pronounced profile which is sometimes the subject of controversial public debates and which is augmented by its location on the water and in the centre of the city as well as by its unusual buildings and thus supplies a memorable image. This surplus of significance requires local representatives and especially the residents to get to grips with the HafenCity and form an opinion about it as a new part of the city. This sort of engagement does not end when the moving vans drive off but is always re-actualised along the lines of the dynamics characteristic of the HafenCity, and plays a large role in allowing residents to achieve 'emotional location attachment' (Reuber 1993: 116) or else to decide to move away from the HafenCity. This form of location attachment, which is not only based on a cost-benefits assessment (rational location attachment) nor on social embeddedness in the location (social location attachment), is based on a particularly high degree of affinity between people and their living environment and normally appears after a long period of living in a given place. It seems that the process of building a relationship to the local context, and with it identification with the neighbourhood (and not just the building where one lives) is accelerated in high-profile locations in the city centre, like the HafenCity, because of their surplus of significance.

This favourable point of departure, secondly, does not by any means lead necessarily to the formation of networks or the development of neighbourliness. Neighbourliness, especially within complex inner-city locations, requires active planning and the well-aimed provision of motivation and also interventions in both the architectural and spatial framework and the developing forms of social interaction. Some examples of such interventions were discussed earlier in the chapter. A concentration of residents set on privacy and with little attachment to the location is not a necessary characteristic of inner-city living. It is, however, a possible development, when, for example, little residential islands rather than differentiated living arrangements are realised and there is too little differentiation of lifestyles and social classes. In such cases, withdrawal into the private realm and avoidance of local life is a foreseeable consequence.

Even after completing actual construction and after the buildings are inhabited, certain forms of support for neighbourliness are needed in places where there is a high pressure on usage and diverse interests in play (and this includes inner-city locations like the HafenCity). Firstly, it is necessary to allow all interested parties to lay claim to public space and thus to ensure a balance of sometimes contradictory interests. Highly commercial interests must not be allowed to dominate public spaces with their presence (eventisation) and influential local parties must be inhibited in the pursuit of their personal interests when they are seeking to block out any usages which could be loud or restrict the presence of *strangers* (people who do not live in the quarter or specific social groups). In high-demand locations like the HafenCity, such processes of negotiation cannot be left to themselves since there is a danger that weaker usages would be crowded out and the intended character of the quarter would be undermined.

What is of particular importance for the development of inner-city neighbourhoods is the achievement of intensive attachment to the location so that local parties see a sufficient reason to focus their social activities in the local arena despite competing alternative options. Local involvement and local networking must thus be stimulated and facilitated, and, equally central, the chance to make a difference and the advantages of the location must constantly be re-articulated. In summary, it is necessary to initiate processes of social mobilisation related to the location. Lively and constructively involved neighbourhood activities cannot be planned down to the finest detail, but neither do they come from nothing. While the independent development of networks is highly important, the planning and pre-structuring and need-oriented support of neighbourliness is essential, even if it is always clear that this must not become a case of the production of set conditions in the sense of social engineering (Etzemüller 2009), but rather it requires the creation of a framework within which independent and increasingly self-regulating social processes with sustainable identification with the local context can arise. Neighbourhoods which are developed, and sometimes professionally supported, within an open-ended perspective are, according to our findings, a new type of social community in inner-city areas. Diverse segments of the population develop a common identity as the heterogeneous residents of a given neighbourhood. Differences become tolerable within smaller social circles working to pursue concrete interests, with whatever degree of friction, and this acts as a productive force in producing the social and cultural attraction of an urban living environment. By this means, neighbourliness loses its traditional image as a mechanism of control and aid and is transformed into a modern impetus for the development of social and cultural capital within urban space.

References

Böhme, G. 2006. *Architektur und Atmosphäre*. München: Fink.

Bruns-Berentelg, J. 2010. HafenCity Hamburg: Öffentliche Stadträume und das Entstehen von Öffentlichkeit, in *HafenCity Hamburg. Neue öffentliche Begegnungsorte zwischen Metropole und Nachbarschaft*, edited by J. Bruns-Berentelg, A. Eisinger, M. Kohler and M. Menzl. Wien and New York: Springer Verlag, 424–55.

Castells, M. 2002. *Die Macht der Identität. Das Informationszeitalter II*. Opladen: Leske + Budrich.

Dangschat, J. 2009. Das Down-Town-Syndrom. Über die Wiederbelebung der Innenstädte – aber zu welchem Preis? *vhw, Forum Wohnen und Stadtentwicklung* 5: 255–7.

Dangschat, J. 2010. Nachbarschaft früher und heute. *vnw Hamburg – Mecklenburg-Vorpommern – Schleswig-Holstein: Engagement für gute Nachbarschaft*, 7–8.

Davidson, M. 2010. 'Love thy Neighbour?' Social Mixing in London's Gentrification Frontiers. *Environment and Planning A* 42(3): 524–44.

Diller, C. 2002. *Zwischen Netzwerk und Institution. Eine Bilanz regionaler Kooperationen in Deutschland*. Opladen: Leske + Budrich.

Etzemüller, T. 2009. *Die Ordnung der Moderne. Social Engineering im 20. Jahrhundert*. Bielefeld: Transcript Verlag.

Feldtkeller, A. 1994. *Die zweckentfremdete Stadt. Wider die Zerstörung des öffentlichen Raums*. Frankfurt am Main and New York: Campus Verlag.

Hoch, H. and Otto, U. 2005. Bürgerschaftliches Engagement und Stadtteilpolitik, in *Handbuch Sozialraum*, edited by F. Kessl. Wiesbaden: VS-Verlag für Sozialwissenschaften, 493–511.

Löw, M. 2001. *Raumsoziologie*. Frankfurt am Main: Suhrkamp Verlag.

Menzl, M. 2007. *Leben in Suburbia – Raumstrukturen und Alltagspraktiken am Rand von Hamburg*. Frankfurt am Main and New York: Campus Verlag.

Menzl, M. 2010. Das Verhältnis von Öffentlichkeit und Privatheit in der HafenCity: ein komplexer Balanceakt, in *HafenCity Hamburg. Neue öffentliche Begegnungsorte zwischen Metropole und Nachbarschaft*, edited by J. Bruns-Berentelg, A. Eisinger, M. Kohler and M. Menzl. Wien and New York: Springer Verlag, 148–65.

Menzl, M., González, T., Breckner, I. and Merbitz, S. 2011. *Wohnen in der HafenCity. Zuzugsmotive, Alltagserfahrungen, nachbarschaftliche Aktivitäten*. Hamburg: HafenCity Hamburg GmbH.

Reuber, P. 1993. *Heimat in der Großstadt. Eine sozialgeographische Studie zu Raumbezug und Entstehung von Ortsbindung am Beispiel Kölns und seiner Stadtviertel*. Köln: Geographisches Institut der Universität zu Köln.

Schroer, M. 2006. *Räume, Orte, Grenzen. Auf dem Weg zu einer Soziologie des Raums*. Frankfurt am Main and New York: Suhrkamp Verlag.

Selle, K. 2004. Öffentliche Räume in der europäischen Stadt – Verfall und Ende oder Wandel und Belebung? Reden und Gegenreden, in *Die europäische Stadt*, edited by W. Siebel. Frankfurt am Main: Suhrkamp Verlag, 131–45.

Simmel, G. 1903/1995. Die Großstädte und das Geistesleben. *Die Großstädte und das Geistesleben. Gesamtausgabe* 7: 116–31.

Spiegel, E. 2003. Stadtplätze als öffentliche Freiräume: Wer nutzt sie wann, wie und warum? In *Was ist los mit den öffentlichen Räumen? Analysen, Positionen, Konzepte*, edited by K. Selle. Dortmund: Dortmunder Vertrieb für Bau- und Planungsliteratur, 175–82.

Statistisches Amt für Hamburg und Schleswig-Holstein. 2010. Struktur der Haushalte in den Hamburger Stadtteilen im September 2009. *Statistik informiert Spezial III/2010* [Online]. Hamburg, Kiel. Available at: www.statistik-nord.de/uploads/tx_standocuments/SI_SPEZIAL_III_2010_01.pdf [accessed: 11 January 2011].

Straus, F. and Höfer, R. 2005. Netzwerk und soziale Projekte, in *Handbuch Sozialraum*, edited by F. Kessl. Wiesbaden: VS-Verlag für Sozialwissenschaften, 471–91.

Trunec, K. 2010. Nachbarschaft und Wohnzufriedenheit. *vnw Hamburg – Mecklenburg-Vorpommern – Schleswig-Holstein: Engagement für gute Nachbarschaft*, 13–15.

Wolfrum, S. 2010. Stadt, Solidarität und Toleranz. *Aus Politik und Zeitgeschichte* 17: 9–15.

Chapter 10

Assessment of the Effects of the Built Environment for the Organisation of Social Processes

Thomas Perry

Urban planning possibilities and limitations are among the important questions addressed in this volume, and particularly in this chapter. The perspective from which I wish to view this is that of an explorer of the living world, who empirically grapples with the connection between people and planning in various applied projects. I am focusing on social and market-oriented tasks in the context of urban planning, re-urbanisation and the future of urban society.

Those who plan should consider the possibility of failure, particularly with projects of the magnitude of those which, like the Hamburg HafenCity, are up for debate in this case. This applies all the more when historical growth (and its auto-corrective powers) is confidently done without, as is the case with the New Downtowns, and instead one relies exclusively on the strategic planning of massive areas and on construction volume. In the decades following the Second World War, the results of such projects the world over were – to put it politely – mixed. The planners' bitter defeats are visible everywhere and add to the spectrum of problems of large cities to this day.

This is not intended as a condemnation of New Downtowns and it is not what this chapter is about. Time will tell. However, these introductory remarks should emphasise how crucial humility is with regard to the dimensions and the importance of the New Downtowns for its users, for the world of finance and the cities vis-à-vis the task that they face. It is important because no one can dispute the abovementioned causal relationship between the way an area is constructed and the social processes, social practices and social life that occur there. Thus city planners can not only do good but they can also thoroughly bungle things. Furthermore, no one can give any concrete examples of how this relation works when, where and under which circumstances and usually the debates on the effects of the built environment on social life end with a furious 'somehow'.

Such blurred certainties can have fatal consequences. On the one hand their lack of focus makes them tend towards banality. On the other hand everybody takes from them according to his liking, twists things the way he wants and creates justification and legitimisation. This state of affairs is all the more regrettable given that the issue at hand is highly important. It addresses both the aspect of living in the neighbourhood – the actual reason for the building projects – and also the market realities, which develop towards whatever type of building. Which buyers or tenants will be interested in it? How will the residence structure develop, and what does that mean for the economic and social sustainability of any given building?

The question will not answer itself, given that the circumstances are too complex. We are living today in a multi-dimensionally layered society of great heterogeneity, which has a central significance for each individual building situation and for the topic in general. This is because modern urban societies are composed of very diverse sub-cultures and living environments, with widely varying orientations and preferences, which have a large impact on communal life. Many of these sub-cultures cannot and will not live with each other. If they did, conflicts would arise which they are trying to avoid. As a result, concentrations of particular groups and segregation arise, as well as images of urban areas and neighbourhoods, selective perception and prejudice, which have a major structuring influence on social processes, our cities and the communal life within them.

In light of this, the scale of claims being raised by the New Downtowns' protagonists is ambitious. Between stage presentation, flagship politics and fleeting events as a form of performative conditioning, the life of a global society is supposed to be developing and the New Downtowns are to serve as its new homestead. All this planning is taking place at the drawing table.[1] And naturally the planners see themselves in a position to deliver. One is faced with the planners' implicit self-perception promising genius but which might equally come to be considered to be naive (or even opportunistic) hubris later on. Which is exactly why the astonished audience has every right to ask some questions.

One should be taken aback by the immanent referencing of the so-called global society. As 'reference and target group' it plays a central part in the design of the New Downtowns. As a living environment it is an innovation which is derived from the dynamics of globalisation. Now, no one will dispute the fact that globalisation exists and that it levels borders, creates supranational cultural elements and generates new transnational environments. But just as undisputed is the fact that this is a comparatively recent process which is always in flux and not yet perceivable in its crystallised form. Hence, from an empirical viewpoint the concept of a global society is extremely wobbly and imprecise, even though it appears to have taken a very precise form as a cliché in many people's minds. But clichés are not sufficient. They might work as concepts but certainly not as target groups. From the perspective of market research and of potential investors the definition of target groups needs to be far more precise. Target groups need to be tangible, describable in precise terms, they need to be real and measurable in size and one has to be able to deal with them in very concrete ways. However, for the time being none of this applies to the global society.

---

1   See also Chapter 1, this volume.

The notion of a global society is also useless as a practical target or reference group because it does not exist as a homogeneous culture or group. Like any modern (and especially post-modern) society it consists of heterogeneous sub-cultures. It is a fatal mistake to speak of *the* society, *the* consumers and *the* citizens in a national context, and it is all the more problematic to speak of a transnational global society in the light of globalisation's constantly changing dynamic. Effectively, the complexity and diversity of contemporary societies is increasing all the time. Globalisation does not resolve this complexity but inflates and drives it. This implies difference, separation, polarisation and heterogeneity of ways of life and identities. Naturally, the group of global elites that one suspects is masterminding the picture of the global society is not excluded from this either.

If this is the case it has to be taken into account at all costs in the planning stages of the New Downtowns. Close attention has to be paid to the differences within the postulated global society as well as to the differences between these groups and the rest of urban society. Only if the differences can be acknowledged and taken into account can the (global and the local) social implications be fully understood as a foundation for the planning stage and be provided for. The concrete planning process of a New Downtown requires more than the consideration of abstract target groups which have been deduced only in theory, and more than the assumption that one knows how to cater for this global society and the cities that accommodate it, for one easily succumbs to the handy but inflated clichés which, in a highly ambitious planning context that tends towards monumentalism and relies on showcasing, are more often than not adapted to the designers' dreams, self-perception and commercial interests rather than born out of an attempt to grapple with real, existing reference groups. It is also all too easy to succumb to self-serving references to projects in Copenhagen, Istanbul, Barcelona, making them serve as empirical confirmation. What is easily overlooked is the fact that these projects have yet to prove whether and how they are viable, which culture they will attract and whether they are more than prestigious business districts which have to be animated artificially after working hours or else will become desolate and alienated from the old town.

The issue of the target and reference groups and their views is therefore undoubtedly relevant to planning. It is also by no means new. In the case of the New Downtowns it is, however, becoming more dramatic as the projects are taking on vast proportions and the positive as well as the negative consequences are therefore all the more serious. In this respect one should consider carefully whether one can afford to gamble or might rather systematically reduce the risks. Let us therefore take a closer look at the practice of target group analysis. One can demonstrate how this is manifested concretely in projects by looking at two basically different planning approaches which one encounters on the real-estate market as well as in other markets and which both engage with the reaction of people to the planned offers.

Goal Orientation

In this form of planning, urban planners, developers, architects and politicians develop a concept which defines and justifies the main points of reference for the construction project. Usually tribute is paid to the quality of the co-existence of more or less well-defined target groups in abstract and normative goals. Then the project is carried out accordingly, or not.

Whether the results are good or bad, relative to the original plan, depends on the perspectives and the interests of the planners. Often enough, obscure goals allow everyone to feel their aims have been achieved. Opinions about the reality of the situation after completion differ significantly especially when they are not unanimously positive and the organizers find themselves under pressure (e.g. due to public opinion).

Demand Orientation

When an area becomes available for development, planners give the demands of target groups a central role. The main question is: for whom are we actually building, and what do those people want? One then tries to build what the target group seems to want. If the situation and the target group are understood correctly, and if one has a bit of luck, everything works out fine. Otherwise, an alternative must be found.

Both perspectives have their advantages and disadvantages. Both also suggest that the effects of the development project on the co-existence of diverse groups can be planned and controlled. One might question whether that is actually the case. We can at least say, however, that the built environment is the context for social processes. To put it more simply: people live there. Every building development has effects on how the people live or can live there, and on which people it attracts or repels. Still, these effects can never be understood separately from their context. And their context is very broad.

Both approaches, therefore, are unfeasible without the intensive engagement with the people whom this is ultimately about. A broad spectrum of methods are available for illuminating essential aspects of such planning with direct reference to and direct contact with the *end user*. This means not only data which lead to conventional examinations of economic or socio-demographic structures. Instead, it means a repertoire of instruments of empirical market and social research, from target-group segmentations to surveys of different kinds to micro-geography, which allow for the small-scale analysis of aspects of living environments. These are instruments which look directly at people in their everyday lives instead of approaching real life via plain description (as in social demography) or indirect, abstract data which are far removed from the social processes mentioned here (as, for example, in data on the economic situation of the city or the region).

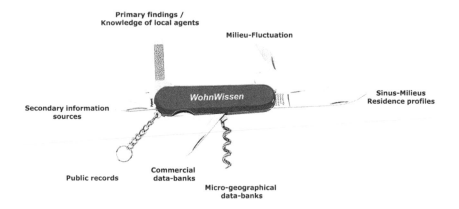

Figure 10.1   WohnWissen – the analysis tool

By observing and analysing who lives where why and with which neighbours, we can gain access to everyday life in constructed space in a very practical and informative way. In an additional step, we can then include constructed space and its conditioning environment (e.g. the local real-estate market). Finally, we must take the history of each state of affairs into consideration. By connecting these points, we gain a fairly good framework within which we can examine the question of the effects of the constructed environment on social processes.

The tools for this undertaking are ready to hand. Since 2002, we, with our partners vhw-Bundesverband für Wohnen und Stadtentwicklung e.V. Berlin and microm-Micromarketing-Systeme und Consult GmbH, Neuss, have been developing and applying a platform (WohnWissen), with which this work can be linked into multiple methodological approaches (Hallenberg and Poddig 2005; Schmal and Wolf 2003).

1. With the social milieus we examine the everyday life of people and dissect their lifestyles and living environments. We analyse, measure and explain what motivates the consumers,[2] why households with the same or similar social-demographic features manifest differing residential demands, and which motives and behaviour patterns guide them, why living somewhere can have very different meanings, which buildings and residential areas are preferred and what significance is attached to the cost-effectiveness of money, among other things. The social milieus are a model for analysing society, consumers and citizens, which has been applied for several decades.

---

2   A lot of publicly available data can be used for that. See for example the Typologie der Wünsche, published by Burda-Verlag, or the Verbraucheranalyse published by Springer-Verlag, both on a yearly basis.

Today it is considered the reigning form of psychographic analysis and offers the know-how, data and knowledge basis which no other comparable model can match. This basis is constantly maintained, updated and expanded, e.g. by annual representative surveys which the vhw does on the topic of residence, and on the basis of which detailed residential profiles of the social milieus have been developed (Appel et al. 2005; Appel and Perry 2005; Beck 2008; Beck and Perry 2007).

2. In order to assess general supply and demand structures at the micro and macro locations, we rely on official large- and small-scale statistics and on available secondary information sources. In this way, reference information can be gained in order to evaluate the market position of an entrepreneurial (partial) portfolio in terms of price, floorspace and qualitative characteristics (Hallenberg 2010).

3. Micro-geographical data, which are available for the whole area down to the level of address and street, and which include a broad range of information (such as data on the type of development, milieu membership, buying power or fluctuation), are connected to entrepreneurial portfolio data. Thus spatial and structural constituent segments can be precisely isolated and analysed with regard to central characteristics of the potential consumers.

4. With the help of records of applications to have mail forwarded by the German postal system which are evaluated systematically by microm, patterns of relocation can be assessed within small-scale structures. By linking this data to further bits of micro-information, the milieu membership of the households which are relocating and/or the old or new type of residence can be analysed. In this way, conclusions can be drawn as to which milieu move into or away from specific regions, specific types of buildings or specific urban areas (see Figure 10.2).

5. The social milieus, as central analytical groups, can be assessed through individual surveys, e.g. in tenant polls, so that very specific interests on the part of a given company or municipality can be taken into account.

6. And, finally, qualitative studies on the basis of milieus can be carried out, exploring important topics for local life and living situations with the local people, e.g. in group discussions or individual interviews. This type of research is of extraordinary importance in pursuing issues of the social processes in a neighbourhood.

Social and lifestyle structures can be described in-depth on the basis of this information, which also provides good data on development structures. One thus gains a qualitative and quantitative framework for the analysis of social processes, but also for the description and interpretation of market relationships. Companies – or other users of this resource, like municipalities – can tailor their strategies and policies much more precisely to meet the demands of the consumers.

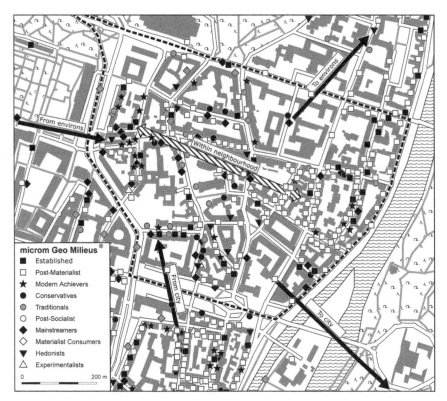

Figure 10.2   Example of micro-geographical data: small-scale milieu fluctuation

Source: Microm Micromarketing-Systeme und Consult GmbH (2008).

To give a brief example: far more members of privileged milieus (established, post-materialist and modern achievers) move out of a gentrified neighbourhood near the city centre to other cities, than move into the same neighbourhood from other cities. The analysis shows that, while the quality of residences on offer fully corresponds to the demands of these milieus, there were too few condominiums available, and those available could not offer competitive prices compared to similar residences in similar neighbourhoods of the neighbouring cities. By reducing this deficit, the attachment of these milieus to the neighbourhood in question could be significantly reinforced.

In this way, a qualitatively and quantitatively differentiated image of moving patterns arises which is developed in connection with small-scale fluctuation data from official statistics and also reflects the inter-relationship which influence the social processes of the area.

Particularly good analyses can be made of places where people are already living. With projects like the HafenCity, the situation is more challenging. There is nothing to analyse yet. One can at least examine the environment and extrapolate from experience gathered in other places.

That is precisely what we did for another client who is responsible for the development of a large space with varied functions within the HafenCity. Retail space, dining facilities and residential buildings all belong to the space in question. Especially with regard to retail space, there is latent competition with the city centre. Given the planned mixed usage and the many other development projects which formed the environment of the HafenCity, it was important to consider which people were to be anticipated as residents and customers in the planned area, and to whom the planning should be oriented. The goal of the analysis was the identification of potential target groups as the basis for a target-group oriented development strategy.

In the process, it was not possible to proceed from an existing environment, since the HafenCity does not yet exist. It was, however, possible to do a qualitative and quantitative analysis of the larger neighbouring environment and the social and commercial life in both the city centre and in the city as a whole. This allowed inductive conclusions to be drawn as to the positioning of the area to be developed which could be matched to the preferences of an empirically identified target group and their preferred social environment and lifestyle.

This type of procedure and the corresponding prognosis of the effects is necessarily based on an empirically grounded foundation in the qualitative and quantitative aspects of social life in the area being developed. This can only be achieved, however, with systematic and regular research on these topics which actively includes the (potential) residents and their perspectives.

In connection with concrete research and advising work, different building blocks can be combined according to requirements. These include elements such as:

- spatial or qualitative segmentation of the object;
- comparative social-space analysis of the area being investigated;
- comparative evaluation of the market rank of available housing according to pricing, space and quality;
- description and evaluation of the tenants according to milieu structure, spending power, age and household composition, among other attributes;
- evaluation of the market context: do supply and demand mesh qualitatively and quantitatively;
- deduction of qualitative prognoses and scenarios: evaluation of the development potential of existing objects or areas depending on alternative development efforts and target-group strategies in the context of general market conditions;
- drafting of an implementation strategy: development of measures and strategies on the basis of qualitative analysis.

In practice, these resources are far too rarely applied. Instead of engaging people directly, experts are called in who either speculate without any empirical basis whatsoever, or state their position on the basis of very abstract, usually superficial data.

The cause of this is a culture of focussing mainly on the demand side of the equation which is still coloured by decades of a supply-oriented real estate market in which demand-orientation was not necessary. Even today, systematic, target-group oriented research on the results of development (for success on the market as well as for local life) is still not an established part of the planning process. In comparison to other economic sectors which depend on the consumer, an expert-ocracy has been established which believes it can manage quite well without direct engagement with the consumer and without primary market research. The real-estate market is thus one of the only consumer markets in Germany which persists in resisting giving central consideration to the consumer. The result is an array of myths, half-truths and wild speculation about trends or the effects of development on social processes which are passed back and forth between experts and passed off as demand-orientation, or which have to serve as a legitimating factor for a supply orientation.

If we are honest, this does not bother many people. What is important at the end of the day is less the effects of the developed residential environment, and more the economic pay-off. Many planners and developers are fond of claiming that the market situations and the value-for-money of the objects on the market determine the later tenancy by up to 60–70%. The effects of the developed area and the perception of the demand side of the equation are very far in the background vis-à-vis such perspectives on the determining factors of tenancy.

However, even on that claim, there is still 30–40% which has to be explained by other factors. These include a large number of aspects which vary according to market, target group, city and type of development. There is no point in surveying them all here. There is no magic formula, universally valid recommendation, or easy answer. Quite the opposite is the case: hard work and implementation of appropriate resources are necessary. The resources needed are actually peanuts compared to the sums spent on other parts of such projects.

The task is certainly not easy. However, in the meantime there is a highly differentiated set of methods with which one can approach the main question of this chapter and the consequences which arise out of it. Still, such resources are initially no more than tools which determine neither the results nor the conclusions drawn from them. Everyone who concerns themselves with this type of research knows how important the prior assumptions and agenda of those using or commissioning the use of such research tools are.

We must thus return to the planners. They are the most essential research context. Their assumptions, goals and perspectives (as in the planning approaches described above) have a decisive effect on the interpretation of the results of research from the very beginning. Their goals and their normative basis must therefore be made explicit from the start, put into concrete terms, and checked for implicit assumptions and contradictions which arise from them. The goals must be related to concrete groups of people: who is actually meant?

In as far as we are concerned with the postulated effects of development on social life, one has a large body of reference material to fall back on. One can refer to other cases, experiences, knowledge and data and use them as a basis for empirically testing the validity of assumptions.

It should also be established whether and how these goals mesh with those of the target group. To do that, the target group must be engaged with directly. This can be achieved through systematic market research, for example with the tools described above. This is no different from what market research does in other markets, and nothing different is what is urgently needed for the real-estate market and for urban planning.

But one thing is not acceptable: one may not remain on the level of normative pleas for a certain goal, garbled expressions and unsubstantiated claims. First and foremost, planners and developers must demonstrate that their plans really do lead to the achievement of their goal. The sad state of the discipline of planning in this country shows that this is not often the case. The results catch up with us again and again. This republic is full of projects which are started with impressive verbal aplomb but end as monuments to the negative effects of development on social processes.

sinus:

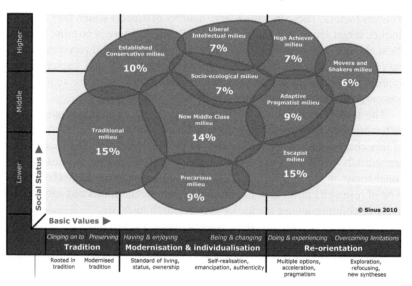

Figure 10.3   The social milieus
Source: Sinus (2010).

Social Milieus: The Approach

The concept of the social milieus has been applied to sociological research for almost three decades. This chapter takes a look at the living environments of our society and is an attempt at an ethnology of daily life. In its segmentation people are therefore grouped together according to their resemblance in terms of lifestyle and ways of living and who share basic values and attitudes to work, family, recreation, money and consumerism. Nowadays it is considered to be a state of the art psychography, providing a solid foundation of knowledge, data and know-how. The categorisation of target groups by Sinus Sociovision is oriented to the analysis of lifestyles in our society. The social milieus group people according to their lifestyles and their perspectives on life. The analysis includes both fundamental issues of worldview and attitudes to aspects of everyday life, like work, family, leisure, money and consumption.

This concept allows a holistic view of people and their environment and thus provides crucial information and assistance for the decision-makers in the areas of planning and marketing in the decision-making process. The concept has been used successfully by enterprises and providers of consumer goods and service providers with their product development and communication since the beginning of the 1980s. Since about 2002 it has been utilised increasingly by local authorities, construction companies and by some developers for research purposes.

The Positioning Model

There are no hard boundaries between the milieus; lifestyles cannot be as precisely delimited as social classes. We call this the fuzziness of everyday life. A basic component of the milieus concept is that there are bridges and points of contact between the milieus.

This tendency to overlap, along with the position of the milieus in society according to social class and basic orientation, are illustrated in Figure 10.3: the higher a milieu is on the diagram, the more advanced the level of education, income and profession; the farther right it is on the diagram, the more modern its basic orientation. Products, markets, media and so on can all be positioned on this 'strategic map'.

With the integration of the social milieus into the most important market media studies and into the AGF/GfK television panel, interesting improvements for media evaluation, beyond product development and marketing, have become possible.

An Overview of the Social Milieus

Socially Privileged Milieus

- Established – 10% of the population, representing the self-confident establishment with an ethic of success, a can-do attitude and a decided taste for the exclusive.
- Post-materialist – 10% of the population. Enlightened liberals: a tolerant attitude, post-materialist values and intellectual interests.
- Modern Achievers – 10% of the population. The young, unconventional highest achievers with intense private and professional lives, preferring to have multiple options and flexibility and enthusiastic about technology.

Traditional Milieus

- Conservatives – 5% of the population. The conscientious German educated class who demand a high standard and are critical of the present and its concept of culture, with a classical humanist education and attitude.
- Traditionals – 14% of the population. Today's elderly post-war generation who want security and order and reject change, deeply rooted in the lower middle class or in traditional working-class culture.
- GDR-nostalgics – 10% of the population. The often frustrated losers of the regime change in East Germany who still treasure the safety and values of the GDR and now feel cheated, they tend towards Prussian virtues and old socialist ideas of justice and solidarity.

Mainstream Milieus

- The bourgeois centre – 15% of the population – the modern centre of society: personal status is important as is the pursuit of established, safe and harmonious relationships and social acceptance.
- Consumer-materialists – 12% of the population. They constitute the modern lower class which follows consumerist values, trying to keep up with the standards of the broad centre but always in danger of slipping into difficult living conditions.

Hedonistic Milieus

- Experimentals – 8% of the population. Extremely individualistic participants and creators of mostly urban life, living in and with the contradictions of modernity with an emphasis on spontaneity and creativity.
- Hedonists – 11% of the population. This is the fun-oriented modern lower class and lower middle class who reject the conventions and expectations of bourgeois society and focus on making the present as pleasant as possible.

References

Appel, C. and Perry, T. 2005. Trendmonitoring im Wohnungsmarkt. *vhw Forum Wohnen* 1: 3–10.

Appel, C., Hallenberg, B., Perry, T. and Poddig, B. 2005. Wohnen in der Sinus-Trendbefragung 2004. *vhw Forum Wohnen* 3: 114–21.

Beck, S. 2008. Lebenswelten von Migranten: Repräsentative Ergebnisse zur Studie Migranten-Milieus. *vhw Forum Wohnen* 6: 287–93.

Beck, S. and Perry, T. 2007. Migranten-Milieus: Erste Erkenntnisse über Lebenswelten und wohnungsmarktspezifische Präferenzen von Personen mit Migrationshintergrund in Deutschland. *vhw Forum Wohnen* 4: 187–95.

Hallenberg, B. 2010. Veränderung der Lebensstile von Senioren. *Raumplanung* 149: 113–15.

Hallenberg, B. and Poddig, B. 2005. Wissen, wer wo wohnt – Das Beratungsangebot WohnWissen. *vhw Forum Wohnen* 4: 212–18.

Schmal, K. and Wolf, A. 2003. Nachfrageorientierte Wohnungspolitik – Weiterentwicklung des Projektes. *vhw Forum Wohnen* 4: 202–9.

Chapter 11

Can Urbanity be Planned?
Comments on the Development of
Public Spaces in the
HafenCity of Hamburg

Claus-C. Wiegandt

Building is not simply the concern of investors and architects; public interest always has a say in the matter. A building should be considered not only in terms of its usefulness, but also in terms of the contribution it makes to life within a city, to the harmony of urban space and to the impression it makes on citizens. All of this calls for a particular 'third party' in the planning and construction process.

Johannes Rau, address on the occasion of the first
Convention on Building Culture (4 April 2003)

Introduction: A New District to Become Urban

In several cities, not only in Germany but also in many other European and North American states, opportunities have been seized throughout the last few years to use derelict inner-city spaces for new large-scale urban building projects. Throughout this process, various public as well as private agents who are responsible for the development of these projects often come up with similar concepts for urban planning. A preferably very detailed and refined mix of usage types plays an important role in these plans, as does the challenging development of public spaces or the discerning blend of existing old buildings, mostly under a preservation order, with appealing and at times spectacular contemporary architecture. Some of these new projects can make use of the additional advantage of being located near water.

One such new large-scale urban building project is the HafenCity of Hamburg. Since the end of the 1990s, a new city district comprising roughly 155 hectares is being developed in the immediate vicinity of Hamburg's inner city. This new district is designed to stand out for its multi-faceted mix of uses as a working, residential, commercial and cultural hot spot. Until recently, the area was reserved mainly for use as a harbour and was therefore principally inaccessible for citizens of Hamburg. Now, however, it is emerging as a unique location within the Hanseatic city which is being built not only for its new inhabitants, but also for all the citizens and visitors of the city and its growing reputation permeates far beyond the city's borders. The numerous reports in the nationwide daily papers and in a wide variety of magazines, including a recent special edition of the *Spiegel* entitled 'Architecture and Design' bear testimony to this. A new inner-city project to the value of circa 5.5 billion euros in private investment and 1.3 billion euros in public investment is currently being built on the old quays with a view to expanding the old inner city of Hamburg by 40% by the year 2025 (current information available on the HafenCity website).

An express wish of the project developer in charge, HafenCity Hamburg Ltd, is to create an entirely urban product (Bruns-Berentelg 2008: 41). This specially founded redevelopment company sees their task as 'defining a new piece of city in terms of urban planning and architecture' and 'creating a model for the development of a European city of the 21st century' (HafenCity Hamburg GmbH 2007: 4). In such a 'highly urban concept' in which objectives for very different dimensions of city planning are being defined, for example with regard to mixed usage, the design and use of public spaces play a special role.

In the master plan for Hamburg's HafenCity, in which the 20-year long planning process was prepared and instigated, stipulations were made that 'particular attention should be paid to the development of public space, particularly with regard to the numerous embankments which serve as locations of both encounter and exchange, possessing a special character all of their own' (HafenCity Hamburg GmbH 2006: 19). In mid 2008, the first central sites were finished. That is why it is time to address the question of how to meet the challenge of using the development of public spaces for the creation of urbanity in Hamburg's HafenCity. How is the approach employed in Hamburg to be judged in relation to other comparable large-scale projects and what lessons can be learned from it?

The aim of this chapter is therefore to reflect on the status of public spaces for the development of urbanity. In this, the main focus lies on how a project development company can influence the preconditions for generating urbanity, which is so central to its purposes. What can those agents who are currently responsible for the construction of Hamburg's HafenCity actually do to generate urbanity in the process of developing public spaces? Is an architecturally appealing design enough to influence or even to develop urban locations with the installation of public spaces, or is there a need for something else? Before these questions can be answered, it is necessary to take a closer look at the terms *urbanity* and *public space* in order to examine the aims, methods and actions of the agents in Hamburg. To what extent the desired urbanity will actually become properly established in the future cannot yet be decisively answered at this moment in time and remains to be seen.

Thoughts on Urbanity: A Myth Comes to Life

The HafenCity of Hamburg is stepping forth with the clear intention to create urbanity in the new city district. If one were to ask the academics what urbanity actually is, it would become clear that this term is decidedly ambiguous and very difficult to put a finger on.

From a somewhat more historical perspective, Walter Siebel has observed urbanity throughout various epochs and describes it as characteristic of the ancient city, of the city of medieval Europe, of the industrial metropolis and the city of the 20th century. In doing so, he refers to the classics of urban sociology, namely Georg Simmel, Louis Wirth, Edgar Salin and Hans Paul Bahrdt, and emphasises the 'dominant difference between city and country and the emancipatory hopes that were associated with this difference' (2000: 264). Various forms of urban lifestyle as distinct from the lifestyle of the rural population play a central role in this context and refer to the dimension of the social co-existence associated with the term urbanity. In this regard, urbanity is understood more as a 'cultural-social form of life and not as an attribute of a unique urban-regional structure' (Sieverts 2001: 32). In other words, urbanity is primarily associated with a 'tolerant, cosmopolitan attitude in inhabitants towards each other as well as towards strangers' (2001: 32).

While working on his dissertation, Thomas Wüst (2004: 46) discovered further dimensions to the term urbanity as a result of analysing the content of almost 200 essays on the subject of urbanity. He thus comes to the conclusion that in the numerous texts he examined, the concept of urbanity not only refers to the specific lifestyles associated with different historical periods, but that the term is also connected with other typical characteristics and criteria of the city. These include aspects of a specific urban anatomy as well as a variety of usages in a city. The inclusion of the constructed and functional municipal structures enlarges here on the somewhat urban-sociological understanding of urbanity and allows it to become a key term in the debate on city planning.

Another distinction that should be considered while employing the term urbanity is that of the actual state or process of the phenomenon itself. This reveals that urbanity is seen both as a process of development as well as a result of development. Ultimately, urbanity is described both as a general phenomenon which can appear just about anywhere and as a phenomenon which is restricted to very particular urban places.

What also becomes clear in Wüst's content analysis is that the term urbanity carries primarily positive connotations. This ties in with a frequently normative use of the term. Urbanity is often understood as a 'melting pot for heterogeneous wishful thinking, powerful interests and normative attributions'. This explains why, in contemporary society, the term urbanity is evocative of 'images of the good and beautiful city life' (Wüst 2004: 44).

Thus, the concept of urbanity is currently acquiring the character of a myth which is in danger of becoming reduced to nothing but a catchword (Wüst 2004: 185). The atmosphere of worldliness, of cosmopolitan thinking and tolerance, of intellectual flexibility and curiosity (Sieverts 2001: 32) is associated with a particular image of the city as having 'a strong and vibrant street life'. This is why Wüst is in favour of expressly naming the specific requirements for an urban development policy which can be derived from the discussion on urbanity with a view to providing a true starting point for an urban policy.

In the case of Hamburg's HafenCity, the term urbanity is also used in a variety of contexts. For example, urbanity is seen as an objective which paraphrases specific spatial structures such as mixture and density on the one hand, and urban forms of co-existence on the other. In other words, urbanity is understood as an essential aspect of city planning which should, in future, become a distinguishing feature of the new city district. This type of urbanity should become a part of the new city district, should be consciously brought about, should be planned. In order to achieve this, various concepts for city planning should be purposefully employed. The design of public spaces is an important consideration in this regard.

Thoughts on Public Space: 'The Decision Will Be Made on the Pitch'

The Ruin or the Renaissance of Public Space

In current academic debate as well as in on-site city planning, public space enjoys a privileged place. From one perspective, public spaces count as important sites which contribute significantly to the urbanity of a city, which is why they are even referred to as the 'basic law of the city'. 'A city's integrity, which is personified in its public space, must remain sacrosanct!' With these words, Thomas Sieverts verily defined the role of public spaces. From another perspective, public spaces are places which are traditionally moulded by the public hand. The potential here for influencing the actual material-physical form is generally greater than in relation to structural engineering projects, which are mostly in the hands of private agents.

In the last number of years, the topic of public spaces has become increasingly important in the area of spatial science (e.g. BBR 2003; Selle 2002). The concern that public space will be threatened by the growing construction of large shopping malls, in which access is controlled and undesirable parts of the population are excluded, has been voiced on a number of occasions and has become the basis for a public debate on the topic. Culturally pessimistic keywords such as 'the ruin' and even 'the end' of public space, as well as 'functional loss' or 'privatisation' are central in this debate.

At the same time, however, hopes have been expressed that public space will continue once again to contribute more to the revival of inner cities. The keywords of 'festivals' and 'events' are associated with this point of view. In our current-day adventure society (Schulze 1992), public spaces are gaining more and more significance and are becoming places which provide a 'backdrop for large cultural events and festival-like productions' (Selle 2002: 60). In Jan Gehl's contribution to this volume, mention is made of the fact that the time spent in public spaces has evolved from a forced necessity in the time of industrialisation to a voluntary act in more recent times, whereby increasing importance is attached to the individual's contribution within these spaces. It is certainly necessary to distinguish here between car-free pedestrian zones in the central inner cities along with parks and green spaces on the one hand, and the frequently dominant traffic zones in which the car dictates the use of public space, on the other. This is where the transit function of streets often restricts the time allowed for other purposes, thereby limiting the contribution public spaces make towards urbanity.

Hybrid Relations instead of Public Spaces?

On closer consideration of public spaces, certain distinctions presented recently by Klaus Selle and his colleagues can prove quite helpful. For instance, this research group from Aachen distinguishes between four levels. These include the 'production of space', the 'assignation of property rights', the 'regulation of use' as well as the 'usefulness and the social character of the space' (Selle 2002: 39). In all four dimensions, we find ourselves confronted with the conflicting interests of private and public agents. This is why, in relation to any one particular space, the most diverse constellations of these four dimensions of production, legality, regulation and usability are conceivable. Such a space might actually be under the ownership of private agents, but on account of its constant accessibility, it possesses the character of a public space for its users. That is why, for instance, it comes as a surprise for some observers that the large plazas in front of railway stations are mainly property of the German Rail and that therefore, from a legal point of view, they are not actually considered public spaces. It was not until the reconstruction of such station forecourts in Cologne and Hannover that the question of ownership and, with that, of public rights was actually discussed in the last few years (for Cologne Brzenczek and Wiegandt 2008, for Hannover von Seggern and Ohrt 2003).

Thus a dichotomy between public and private spaces is an oversimplification in itself. In their research project, Ulrich Berding, Juliane Pegels and Klaus Selle therefore address the tension between private and public interests concerning inner-city spaces and verify several so-called hybrid spaces which, with regard to their design, their regulation and also their maintenance have been strongly shaped by private and public agents. As a result, positive as well as negative consequences arise for these spaces from the interplay between communal and non-communal agents. In this way, new publicly useable spaces have latterly been created or upgraded with the result of having somewhat enhanced the image of the respective location. On the other hand, however, a negative impact on aesthetic qualities has been ascertained (Berding et al. 2007). The aspect of aesthetics addresses a third subject area which has assumed an important place in recent discussions surrounding public spaces.

Public Spaces and the Aesthetic Configuration of City Planning

The several attempts to transform and restyle public spaces which can currently be observed in inner cities as well as other city districts are an integral part of the debate on the aesthetic configuration of city planning. After having met with the 'basic needs' required during the age of post-modernity, the aesthetic mores are starting to outstrip bare functionality as an essential feature of the city of today (Böhme 2006: 7). This is particularly valid in the case of architecture and urban planning. Aesthetic needs, which the philosopher Gernot Böhme paraphrased as 'the need to adorn oneself, the need to be seen, the need to make an appearance and the need to raise one's standard of living with the help of fixtures and fittings' (Böhme 2006: 10), correspond with the manifold upgrading techniques employed in public spaces (Brzenczek and Wiegandt 2008).

It is especially the public spaces in our key cities which not only function as traffic zones, but also provide places where people can act out their lives and design a new lifestyle all of their own. The transformation of public spaces in numerous German cities in recent years equally accommodate experiences which embody a part of urban life, as is the case in the restyling of public spaces in new city districts such as Hamburg's HafenCity. In so doing, they are becoming involved in a growing culturalisation of urban policies which are gaining significance in the context of international urban competition (Häußermann et al. 2008: 248).

Strategies in Hamburg's HafenCity –
Public Spaces as a Contribution to Urbanity

In the development aims for Hamburg's HafenCity, public spaces play a particularly important role with a view to achieving the highest level of urbanity possible in the new districts. Solely with regard to quantity, the public spaces in Hamburg's HafenCity will hold more significance than Hamburg's historical inner city. With regard to quality, special attention is being paid to the design of public spaces.

Five Types of Public Spaces in the HafenCity

In future, there will be different types of public spaces throughout the entire HafenCity and they will be distinguished in terms of their respective significance for the district:

- The first type of public space, namely the *waterfront promenades*, distinguish themselves by virtue of their new and somewhat linear quality as well as their immediate proximity to the water. At the end of these promenades, there are terraced plazas which also make use of their proximity to water in order to entice people to linger in the HafenCity. Compared with the promenades, they are more specifically designed as a place in which to spend time. Both the waterfront promenades and the terraced plazas at either end of the harbour basin fulfil an important function within the district in that they offer inhabitants and visitors alike a particularly attractive place to spend time.
- A second type, which the project developers devoted particular attention to, are the smaller district squares between the residential areas. These squares distinguish themselves from the waterfront promenades in terms of their location and size and can serve as a meeting point for the future inhabitants in the new district. Special demands regarding design were also made on these squares.
- Small *green* parks represent a third type of space being developed within the HafenCity. Up until the middle of 2008, these areas were not fully finished, but they continue to constitute a further important element in the network of public spaces.
- A fourth type of public space is the planned Überseeboulevard. This is a pedestrian zone which is to emerge in the Überseequartier and will comprise an open shopping and business area. In place of an originally planned indoor shopping mall, the Überseequartier has been developed as a central location in the HafenCity since September 2007. The Überseeboulevard is a clear example of a hybrid space (Berding et al. 2007). From a legal point of view, the Überseeboulevard will be a privately owned space for which public access rights will be regulated by means of a detailed contractual agreement with the private investor. This central area of the HafenCity will be completed by the end of 2011.
- Ultimately, the *classic* street system of the new district together with its pavements also belongs to the network of public spaces which weave their way throughout the HafenCity. In contrast to the first four types, this particular type of public space has been designated for traffic and is therefore significantly different from the other types with regard to its use.

By the middle of 2008, the first set of public spaces was completed in a significant part of the HafenCity. Not all of the above-mentioned types, however, are to be found. A place of prominence has been occupied since 2007 by the Marco-Polo-Terraces at Sandtorhafen, the Magellan-Terraces at Grasbrookhafen and the Dalmannkai Steps. As an ensemble of spectacular urban spaces located by the water's edge (HafenCity Hamburg GmbH 2007: 6), the new waterfront development should lend a mark of distinction to the urban planning of the new district. Specific requirements for the design of these waterfront zones result from the high water levels. Several properties which were built in the HafenCity have been raised by three metres for reasons of flood protection so that the design of the embankments on either end of the harbour basin stands out on account of the steps which span the difference in height between the ground floor of the building and the normal water level. In so doing, they provide a rather elegant solution for meeting the specifications of flood protection while at the same time fulfilling the expectations of an attractive location.

The Vasco-da-Gama-Plaza was also completed in the middle of 2008. It is integrated into the stretch of quay between Sandtorhafen and Grasbrookhafen in the new residential area and is equipped with a basketball court as its local square. It is the first square of this kind which will later belong to a network of smaller local squares within the individual districts. While these local squares should enable access to the general public, they should also offer the inhabitants of the new districts meeting points and therefore cater more strongly than the waterfront promenades for the needs of the new resident population in the immediate vicinity. Consequently, these squares will be developed individually while at the same time taking into consideration the neighbouring types of usage.

In creating the specific facility of a basketball court on the Vasco-da-Gama-Plaza, architects and project developers created special options for using the space which are strongly geared towards the local residents. It remains to be seen to what extent social practices will be shaped by the design of such urban structures. Only time will tell whether the proposed idea of ballgames will actually be accepted in this area. It is also conceivable that such a utilisation of public space could, in the long term, lead to disturbance in this quarter with the result that ballgames will be disallowed or that the facilities which enable such games will have to be removed. This is a perfect example of how the HafenCity not only influences the form of public spaces, but on account of specific design, also influences the function of these spaces.

Urban Design for Public Spaces

Hamburg's HafenCity poses particular demands on the design of public spaces. They should attain the highest level of aesthetic quality in order to ensure pleasant conditions for time spent outdoors. This is why there is the deliberate intention to create a 'maritime flair' in the new HafenCity which should invite people to linger in public space. All of this contributes to the overall aim both to enliven public spaces in future and to attract as many people to spend time there as possible. In order to achieve this aim, careful consideration with regard to materials and design elements is taking place. Among these considerations are the wooden decks on the Marco-Polo-Terraces along with the small grassy hillocks in this area, special street lamps designed as harbour cranes, as well as the decoration of the basement walls with brick ornamentation.

In all of this, the project development of the HafenCity is adhering to what is meanwhile a well-established appreciation of urban design which lately became a significant factor for all large-scale reconstruction projects (Bodenschatz 2008: 17):

'Attractive city planning ... should contribute towards creating a lively stage which facilitates the image cultivation of the urban middle classes. Such a stage must be designed in a visual and enjoyable manner'. This means 'a reclaiming and a recreation of public spaces, new uses of spectacular water locations, pedestrian-friendly arrangement of streets and squares, a certain structural density and a mix of uses'. The development of public spaces is not only seen in Hamburg's HafenCity as a vital contribution to an attractive, but also to a most urban city.

In 2006, a two-phase international open space planning competition was carried out for the west part of the HafenCity. The winning submission from a Barcelona office stood out from all the others in that the materials used guaranteed the creation of 'a cohesive urban environment for visitors'. At the same time, it is intended to differentiate between 'the specific characteristics of each individual area by means of architectural design' (HafenCity website). Hence, a distinctive feature in the design of public spaces is a clearly identifiable language of form which lends expression to the area as one complete harmonious unit. As in the case of other large-scale urban redevelopment projects of recent times, the public spaces will not 'be produced as islands, but will be connected with each other ... in a sort of urban labyrinth' (Bodenschatz 2008: 18). This recurrent theme of a harmonious design should allow visitors to experience the new district as a cohesive unit, while at the same time allowing them to distinguish it from other city districts.

In their design of the various individual sub-areas of the waterfront promenades, the architects from Barcelona have since created lying and seating areas of stone modelled on wave-like images, similar to the forms that can be found on the strand promenades in Barcelona. These are proof positive of the project development company's commitment to create exciting public urban spaces which offer as many options for use as possible. In this way, not only aesthetic aspects are included in the consideration of the structural design of the waterfront areas, but also different potentials for use. In other words, the design is not only concerned with aesthetic requirements, but also with the desire to offer the public the broadest possible spectrum of use.

A cobblestone pavement lying adjacent to a smooth concrete surface makes it possible to stroll, to promenade and even to skateboard along the water's edge. The presence of skateboards was, however, rendered impossible by a subsequent reconstruction of the railings attached to the steps. There was a danger that the skateboarders would very quickly have usurped these steps as an exciting territory of their own, thereby endangering or displacing other passers-by in the process. This is why the project development company felt prompted to do up the railings with the help of creative measures in order to restrict the use of skateboards and protect weaker segments of the population. This illustrates how a project developer responds in a regulative manner to specific forms of usurpation in the interest of the entire project and the majority of users.

It is to be expected that the HafenCity will probably become an ideal place of adventure for a section of the population who wish to use the public space for cultivating their image and for living out their own individual lives, while at the same time allowing themselves to be influenced by the carefully designed squares and promenades as well as what takes place there. In a unique fusion of leisure, shopping and cultural facilities with this beautifully formed public space, a new understanding of what *public* actually means is becoming apparent for a large number of the public spaces in the HafenCity. There is, however, a palpable danger that this new understanding will imply that particular segments of the population who are less interested in seeking adventure (Schulze 1992: 14) will find such spaces unappealing and will therefore not frequent them.

On the one hand, this poses the question as to what extent these new public spaces will actually be adopted and accepted in future by all the citizens of Hamburg or whether a new form of segregation will emerge. There is a need for empirical research in this regard. Without it being the conscious intention of the decision-makers in Hamburg's HafenCity, there is a suspicion that particular segments of the population will not visit these places. As a result, a new form of what is perceived as *public* could become established, a form which would deviate greatly from what many urban sociologists advocate. It would be a form which understands the public sphere as a place of anonymity, of stylised behaviour, of pomposity, aloofness, indifference and cosmopolitan intellectuality (Siebel 2004: 14). On the other hand, this leads to the question as to whether such new concepts of *public*, which are not only hinted at in Hamburg's HafenCity but are also showing themselves in Berlin at the Potsdamer Platz or in Oberhausen on CentrO's promenade, actually differentiate the sphere of public spaces in our cities and that therefore our classical sociological understanding of public spaces to date should be reconsidered.

The Conclusion: On the Planability of Urbanity

A lot of effort is being invested in the design of the HafenCity's public spaces in order to secure the best preconditions for a usage profile that will meet with the ideal of the European city. In this regard, design concepts are being developed with the help of various instruments employed in the current practice of urban planning. As with all large-scale projects that are developed within a relatively short space of time, there is a certain risk that, while the design poured from a particular mould might be considered highly attractive for a particular period of time, such public spaces which emerged in accordance with uniform specifications will lose a lot of their charm. In this, the feeling of naturally arising spaces, which grows with time in a process of constant adaptation, can be missing. That is why it is an important task of the redevelopment company to allow for such long-term adaptations to the contemporary urban design.

Secondly, the HafenCity should be home to a life of as much variety and colour as possible, facilitating many different interests. A whole spectrum of users should be satisfied, the property owners in the district should be able to avail themselves of good options for utilisation and the city of Hamburg should benefit from a cutting edge in competition with other big cities. The public spaces in Hamburg's HafenCity are therefore conceived as beautiful, clean and safe places. With considerable commitment in terms of planning, places are being created and marketed which are attractive for a large part of society and which will most probably meet with the approval of other social groups. What is more, exclusive public spaces created in this way, and also in accordance with the whole urban concept of a sophisticated mix of uses in the urban city, will meet with fewer negative responses from society. With regard to the ideal notion of what the HafenCity should be, we are looking at an 'image of a "dry-cleaned" urbanity' (Sieverts 2001: 33).

This is where extremely ambivalent criticism sets in: places of an individual and higher quality are emerging which are unable to meet with all the requirements of an urban location, but instead with the desired urbanity of a large majority. For there is the tendency that financially weaker segments of the population feel excluded by the elaborate design of public spaces and therefore do not warm to them. A segregation in public space is therefore to be expected, although this needs to be empirically tested. In this regard, there remains nothing more than an academic cultural criticism which insists on being confronted with the negative aspects of urban life that play themselves out on the city's stage: with the misery of drug-addicts, the poverty of the homeless and also with the insecurities that the city holds ready (Häußermann et al. 2008: 307). Such 'appeals for a heroic urbanity' (Häußermann et al. 2008: 307), however, no longer seem capable of winning a majority.

References

BBR, Bundesamt für Bauwesen und Raumordnung (Hrsg.) 2003. Öffentlicher Raum und Stadtgestalt. *Informationen zur Raumentwicklung* 1(2): 1–110.

Berding, U., Perenthaler, B. and Selle, K. 2007. Öffentlich nutzbar – aber nicht öffentliches Eigentum. Beobachtungen zum Alltag von Stadträumen im Schnittbereich öffentlicher und privater Interessen, in: *Shopping Malls. Interdisziplinäre Betrachtungen eines neuen Raumtyps*, edited by J. Wehrheim. Wiesbaden: VS-Verlag, 95–117.

Bodenschatz, H. 2008. Perspektiven des Stadtumbaus, in *Großstädte von morgen. Internationale Strategien des Stadtumbaus*, edited by H. Bodenschatz and U. Laible. Berlin: Braun, 9–23.

Böhme, G. 2006: *Architektur und Atmosphäre*. München: Fink.

Bruns-Berentelg, J. 2008. Das Verhältnis von Stadt und Fluss neu definieren: Die Hafen City Hamburg. Standort-Gespräch mit U. Bauer. *Standort – Zeitschrift für Angewandte Geographie* 32(2): 40–44.

Brzenczek, K. and Wiegandt, C.-C. 2008. Einheit und Vielfalt? Von Akteuren und Instrumenten bei der Neugestaltung innerstädtischer Plätze. *Die alte Stadt* 35(2): 372–85.

HafenCity Hamburg GmbH. 2006. *HafenCity Hamburg. Der Masterplan* [Online]. Available at: www.hafencity.com/upload/files/files/z_en_broschueren_19_Masterplan_end.pdf [accessed: 11 January 2011].

HafenCity Hamburg GmbH. 2007. *HafenCity Hamburg Projekte. Einblicke in die aktuellen Entwicklungen.* Heft 8. Hamburg.

Häußermann, H., Läpple, D. and Siebel, W. 2008. *Stadtpolitik.* Frankfurt am Main: Suhrkamp Verlag.

Schulze, G. 1992. *Die Erlebnisgesellschaft. Kultursoziologie der Gegenwart.* Frankfurt am Main and New York: Campus Verlag.

Selle, K. 2002. *Was ist los mit den öffentlichen Räumen? Analysen, Positionen, Konzepte.* Dortmund: Dortmunder Vertrieb für Bau- und Planungsliteratur.

Siebel, W. 2000. Urbanität, in *Großstadt. Soziologische Stichworte*, edited by H. Häußermann. Opladen: Leske + Budrich, 264–72.

Siebel, W. 2004. *Die europäische Stadt.* Frankfurt am Main: Suhrkamp Verlag.

Sieverts, T. 2001. *Zwischenstadt zwischen Ort und Welt, Raum und Zeit, Stadt und Land.* Gütersloh and Berlin: Bertelsmann Fachzeitschriften.

Von Seggern, H. and Ohrt, T. 2003. Lehrstück und Streitfall: Ernst-August-Platz: Bahnhofsvorplatz in Hannover. *Informationen zur Raumentwicklung* 1(2): 69–75.

Wüst, T. 2004. *Urbanität. Ein Mythos und sein Potential.* Wiesbaden: VS-Verlag für Sozialwissenschaften.

Chapter 12

The Virtue of Diversity

Rolf Lindner

The city is a place which one can no longer contemplate leaving because it is the most powerful expression of life.

Paul Nizon

The Growing Valuation of Diversity

Diversity has recently become a new keyword in the areas of cultural and social studies. This does not come as a surprise given the changes which have come with the process of globalisation. Due to global mobility and migration, diversity has become a characteristic of post-modernity. The level of importance that has become attached to diversity as a socially established fact can be seen in the growing significance of so-called Diversity Management as a key feature of Business Management. The aim of Diversity Management lies in developing strategies, programmes and measures 'so that the potential advantages of diversity are maximized while its potential disadvantages are minimized', as Taylor Cox (1993: 11), Professor of Human Resource Management at the University of Michigan Business School, noted with the optimism typical of specialists in organisational studies. The field of Diversity Studies has grown beyond Management Studies to encompass a more integrating, inter- or transdisciplinary direction of research, which is concerned with questions of diversity and difference in all social sectors, including economics, politics and education. Within a German-speaking context, Diversity Studies, which subsumes the areas of Gender Studies, Migration Research, Intercultural Studies, Postcolonial Studies and much more, finds itself in a nascent state. In the summer of 2005, a research network entitled 'Diversity Studies' was called into life at the Free University Berlin. In 2006, the Centre for Diversity Studies (Cedis) was founded as an inter-departmental research and teaching group at the University of Cologne. All of these studies share the same perspective, namely that diversity is not merely a problem from an administrative, planning and political point of view, but, more importantly, it also presents an opportunity, or in economic terms, a resource, this being: diversity as a virtue.

Diversity is, however, not a new phenomenon; indeed it has always enjoyed a privileged place which offers an ideal context for the field of Diversity Studies, i.e. the city. That which we emphatically refer to as urbanity is indeed Diversity Management in a generic sense: it embodies the city, the *polis*, as a forum for opinions and a place for encounters.

In his reflections on urban society, the Swedish social anthropologist Ulf Hannerz (1980: 118) observes the following: 'Running into people for whom one was not looking, or witnessing scenes for which one was not prepared, may be neither efficient nor always pleasant, but it may have its own personal, social, and cultural consequences'. What goes hand in hand with these consequences is the increased likelihood of discovering 'one thing by chance when one was looking for something else' (Hannerz 1980). Serendipity is the name given to this frequently underestimated concept in academic research. It is no coincidence that serendipity features as a theme in metropolitan films; there is even a film of the same name set in New York and released in 2001 with Kate Beckinsale and John Cusack in the leading roles. As portrayed by various urban legends, the metropolis is a *privileged place* not only for such chance discoveries, but also for chance meetings which give rise to new contacts and networks. Dieter Laepple is therefore justified in referring to the city as a random generator.

Diversity and Urbanity

'The City', writes urban sociologist Lyn Lofland, 'is not like other kinds of places. The City, because of its size, is the locus of a peculiar social situation: the people to be found within its boundaries at any given moment know nothing personally about the vast majority of others with whom they share this space' (1973: 3). In Lofland's eyes, the city is a world of strangers: 'To live in a city is, among many other things, to live surrounded by large numbers of persons whom one does not know. To experience the city is, among many other things, to experience anonymity. To cope with the city is, among many other things, to cope with strangers' (Lofland 1973: ixf.). It is indeed a world of its own which has become the subject of legend in city narratives of various kinds, be that novels, poems, but especially films and songs ('Stranger in the Night'). If we follow Lofland's line of thought that the city is a world of strangers, we come to the conclusion that xenophobia is synonymous with hostility towards cities (a point which is actually endorsed by the history of hostility towards cities, which has always been associated with resentment or fear of what is strange or unknown).

If there is a keyword that captures the fascination, the atmosphere and the essence of the city, then it is diversity. Diversity is an inexhaustible urban resource and encompasses diversity in place of origin, in lifestyle and in professional careers, to name but a few. This diversity, however, is never stable. As the cultural anthropologist A.L. Kroeber stressed, it is in fact characterised by changing trends 'not only of dress but of fads, novelties, amusements, and the fleeting popularity of persons as well as things' (1948: 283). It is this truly multifaceted diversity and the astonishing variations on life's themes that, in the words of Gunther Barth (1982), are virtually embedded in the urban landscape and which make the city attractive because it is never boring. Indeed, it is the city which satisfies our hunger for variety, characterising it down to its very essence as a theatre of opportunity with its own inherent structures of chance and individual fields of experiment, manifestly displaying that the differences between human beings represent the elixir of life in any modern metropolis. And last but not least, it is diversity which is reflected in and through all urban media.

Vaudeville theatre, the most popular and naturally urban entertainment attraction in North American cities between 1880 and 1930, can serve here as an ideal case in point. Even the name itself, a slightly bastardised form of 'voix de ville' (voice of the town), depicts the symbiotic relationship between varieté theatre, a theatre for variety and amusement, and the city. The vaudeville show dramatised the diversity of urban life as much through its chosen themes as through its variety of acts, which included, among others, acrobats, comedians, dancers, clowns, jugglers, musicians and impersonators, the very definition of an urban artist. It is no coincidence that the term impersonator refers in English to both an imitator and an imposter; in other words a *con(fidential) man*, the ultimate urban criminal. Due to the pace of the acts that reminded the audience of the city, the vaudeville show was captivating. The same can be said of the urban press around 1900, which quite literally gave voice to the town. The daily newspaper symbolises a substantial portion of what seems to characterise a city. Karl Christian Führer writes: 'It stands for pace, restlessness and haste, for the sensationalism that overshadows yesterday's news, thereby embodying all that is transient, exchangeable and in' (2006: 106). The very rhythm of the city is reflected in the publication of newspapers (morning, midday and evening issues as well as Sunday papers), in how they are distributed (namely using street vendors) and in how they are presented. Even the presentation of extremely diverse news items on the front page itself is, as Philip Fisher (1975) emphasised, a form of (printed) city life. It is not only due to the fact that the city papers themselves are a part of the never-ending deluge of impressions, but also owing to their understanding of news as novelty that enables them to stimulate a new supply on a daily basis. Seen from this viewpoint, the urban press is not merely an expression of, but also serves to promote, intensification in life within the modern city. Ultimately, the city itself provides its own medium of intensification by drawing on a fast and continuous exchange of external and internal impressions, which is just what Georg Simmel (1903) declared at the beginning of the 20th century to be typical of a city's nature.

Due to its intricately interwoven scenes, New York was considered the archetypical city of diversity at the beginning of the 20th century: a true variety show. John Reed, who worked as a journalist at the time, perceived New York as a magical city in which everything and anything could be found:

I wandered about the streets from the soaring imperial towers of downtown, along the East River docks, smelling of spices and the clipper ships of the past, through the swarming East Side – alien towns within towns – where the smoky flare of miles of clamorous push carts made a splendor of shabby streets ... I knew Chinatown, and Little Italy, and the quarter of the Syrians; the marionette theater, Sharkey's and McSorley's saloons, the Bowery lodging houses and the places where the tramps gathered; the Haymarket, the German Village, and all the dives of the Tenderloin ... The girls that walk the streets were friends of mine, and the drunken sailors off ships new-come from the world's end, and the Spanish longshoremen down on West Street. I found wonderful obscure restaurants where I could try the foods of all nations. I knew where to procure drugs, where to go to hire a man to kill an enemy, how to get into gambling dens and secret dance halls. I knew well the parks, and the streets of palaces, the theatres and hotels; the ugly growth of the city spreading like a disease, the decrepit places whence life was ebbing, and the squares and streets where an old, beautiful leisurely existence was drowned in the mounting roar of the slums. I knew Washington Square, and the artists and writers, the near-Bohemians, the radicals. I went to gangsters' balls at Tammany Hall, on excursions of the Tim Sullivan Association, to Coney Island on hot summer nights ... Within a block of my house was all the adventure of the world; within a mile was every foreign country. (Cited in Stuart 1972: 16–17)

John Reed's somewhat restless and breathless depiction of New York as a variety show compiled of urban scenes dating back to 1911 is not an isolated account. We can find similar contemporary descriptions of New York in Theodore Dreiser's collection of newspaper sketches entitled 'The Color of a Great City'. In Dreiser's eyes, the marvellous nature of the city lay within its diversity, which expressed itself in contradictions. According to the Chicago-born urban sociologist of the 1920s, the city represented a mosaic of microcosms which invited one to partake in a fascinating experiment of moving within worlds lying adjacent to each other, but otherwise entirely separate entities.

It was in this multiplicity of worlds that the city planner Karl Schwarz saw the distinctive features of cities as places where a stranger 'can rediscover his own world because it [the city] presents the possibility of all imaginable worlds' (1987). In his appellative essay, which was one of the first to define the meaning of cities in terms of their cultural function, Schwarz identifies the specific qualities of a city which reveal its fundamentally irreplaceable quality: *diversity and contradiction*, i.e. the variously contradictory parallel nature of social situations, of human beings hailing from different social and national origins, meshed with the tangible temporal depth, of the synchronicity of all that is out of sync; *hot spots*, meaning the self-perpetuating phenomenon of revelatory and temporary urban spaces as secret places of social experiments and as a platform for non-conformists and the under-privileged to live and develop, *niches* within a world in the grip of functionality and economy; *stages*, meaning public space as a place for experimentation and presentation of individual lives gathered into collective projections: a performance space; and finally, *refinement*, meaning the consciously enhanced artificiality of the city as an entertainment park, along with its accumulation of attractions, its animation of illusions, which taken together should give inhabitants the feeling that they are in a world characterised by the unlimited possibilities of the human imagination (Schwarz 1987).

Diversity in German Cities

A few years ago, the Berlin-based tabloid *Bild Zeitung* printed the headline 'Berlin's 25 best faith healers' on its front page. Alongside its initial intention to inform the reader that there must be many more than merely 25 faith healers in Berlin, this headline launched the first in a series of articles about spiritual healers and mystics, diviners and magicians, arcanic guides, shamanistic fortune-tellers and Brazilian druids, promising to provide the reader with information on the quality, price and methods of each healer. Beside mediums and miracle workers, groups dealing in esotericism, the occult and spiritualism are principally urban phenomena. If we call to mind the prophets of inflation who populated the streets of Berlin in the 1920s, they serve as the perfect example for the fact that the city provides a home for what Georg Simmel terms the most 'tendentious oddities'. What is unique to the peculiarities of the city is the fact that the strangest idiosyncrasies, abilities and knowledge can be turned into a profession. This is a particularly suitable example to illustrate the two fundamental processes that define the culture of the metropolis: the process of inner differentiation which is articulated by the professionalisation of fortune-telling and healing, and in general by the entire spectrum of skills, along with the process of networking which is represented here by the establishment of various groups (Lindner 2005). The potential capacity within cities can be seen most expressly in terms of the creativity that is set free in relation to the invention of new careers, new services and new professional branches. The journalist Tobias Timm wrote in an article that anybody in Berlin can think up a new job description (it is no coincidence that the catchphrase 'digital Bohemian' was coined there). However, I presume that creative people in Hamburg are not lagging behind in this respect given that the same development is being advanced in the trend towards self-employment. From the perspective of Robert Park, a student of Simmel and founder of the Chicago School of Urban Sociology, the city represents a human workshop in which the process of dividing labour is carried out in the most intricate and at times seemingly bizarre manner. Hence, it is no surprise that within the context of his famous 'Suggestions for the Examination of Human Behaviour in Urban Milieus', he specifically proposed a list of career types which he considered worthy of investigation. He writes of the tendency in cities of every activity, be it even that of begging, to assume the character of a profession (Park 1915). It is beyond a shadow of a doubt that this is truer now more than ever: one-man shows, funeral speakers, casting agencies, knowledge designers, erotic photographers, costume rental, ghost writers, hypnotists, intimate jewellery, condom service, products for left-handers, flat-sharing agencies, note-writing service, oriental dance (including a speciality shop for high-class belly dancing costumes and accessories), pantomime, quads, diviners, strip-o-grams (in police uniform, as a pizza delivery man or a postman), taxidermist (also for house pets), translators, gilders, fortune-tellers, yogis, magicians. Berlin's as well as Hamburg's current 'Yellow Pages' prove themselves to be a veritable treasure trove of services and business enterprises which could only survive in a city. Moreover, they bluntly reveal the metropolis's characteristic synchronicity of all that is out of sync. In Hamburg's business directory, woodworm control (*Holzwurmbekämpfung*) is followed immediately by homepage design (*Homepageerstellung*). It is these apparent anachronisms that hint at the unlimited possibilities of the city. Indeed, skills which seem to be on the brink of extinction survive in the city. In Paris, for example, this can be seen in the production of a wide range of small accessories associated with *haute couture*, including such fancy goods such as *passements*, silk flowers, embroidery, etc. In the case of Hamburg, we need only consider the many skills associated with commodities necessary for boat and ship building, for example master upholsterers, rope makers, net makers, etc.

Naturally, in order to make far-fetched plans, it is necessary that like-minded people come together, as we saw from the example of groups of esoterics and occultists in Berlin. This illustrates how networking forms the second fundamental process which constitutes the city as an arena of possibilities. Ulf Hannerz identified a further characteristic in the city's accessibility to diversity:

> What may be latent or barely visible concerns of one or a few individuals in a smaller community … may be amplified when many like-minded people are about … It is in the bigger city, usually, that one finds not just the single pianist but a musicians' occupational culture, not one quiet political dissident but a sect or a movement organized around an ideology, not a lone homosexual but gay culture. (Hannerz 1980: 115)

The city functions here as a cultural catalyst: it is only through networking that it is possible to translate the private status of pianist, dissident or homosexual into a cultural statement.

A third fundamental process alongside differentiation and networking is spatial density or clustering. This refers to the entire array of spatially and culturally differentiated spheres of working, living and entertainment. Sociologists in Chicago have come up with terms such as *natural areas, social worlds* and *moral regions* to describe this phenomenon. They speak of 'cities within a city': urban villages, ethnic enclaves, cultural milieus and sub-cultural niches. Mono-structural service and business enterprises were established even in the early stages of urban history. Examples of this can be seen in the old manufacturing district in Mitte in Berlin and particularly on and around the Hausvogteiplatz, which was developed as early as the 1840s. Further examples can be seen in the old newspaper district in Friedrichstadt, the centre of Berlin's printing and press industry during the Imperial Era and the time of the Weimar Republic, a district which is currently experiencing a mild renaissance. And last but not least, the Siemens City which emerged around 1900 and was referred to in the 1920s as an 'electronic metropolis': a true 'city within a city' with all the necessary infrastructural, communal and social facilities required by the local community such as kindergartens, primary schools, shops, churches, etc. Nowadays, the gallery district surrounding Augustus Strasse, the design cluster around Neue Schönhauser Strasse, the new media district known as Media-Spree, and the science city of Adlershof stand in the light of Berlin's public eye. Similar mono-structural and mono-cultural urban areas were established in Hamburg at the end of the 19th and beginning of the 20th century. Comparable with the area around Koch Street in Berlin, a newspaper district was developed in Hamburg during the 1920s which comprised what at that time could be considered an indispensable spatial fusion of editorial and printing departments in the area surrounding the Gänsemarkt. Karl Christian Führer reveals in his essay on 'Urban Space and Mass Media' that the publishing houses of the 'Hamburger Anzeiger', the 'Fremdenblatt', the 'sozialdemokratisches Hamburger Echo', the communist 'Hamburger Volkszeitung' and the 'Hamburger Neueste Nachrichten' were all located within a radius of less than 100 metres. What is more significant within our context, however, are areas which were characteristic of the city in a special sense. Examples for this can be found in the Speicherstadt, an off-limits enclave which constitutes a true city within the city, and the Kontorhaus district which accommodated the spatial needs of Hamburg's merchants and traders. Spatial clusters are particularly revealing when it comes to the continuity and change of urban economy and culture. From this point of view, the HafenCity is most definitely embedded within Hamburg's history as a city, carrying forward the continuity of maritime Hanseatic areas into post-modernity. From this perspective, the HafenCity appears to me to be an attempt to duplicate the city of Hamburg in an idealised manner, in other words in a manner which surpasses elements of reality. In so doing, it attempts to create a type of flagship store for the city not dissimilar to the ambitions of the Potsdamer Platz in Berlin, which represents a landmark not restricted to one architectonic monument, but encompassing a whole district. That this is not only a bold endeavour, but also one that goes hand in hand with hubris, seems quite obvious to me.

Diversity in the HafenCity

It appears that absolutely everything was considered during the planning of the HafenCity, but maybe therein lies the inherent problem of a master plan. It was a plan which comprised a mixture of multiple uses accompanied by diversity in order to avoid office wasteland on the one hand and dormitory cities on the other and which followed the ideal of the detailed mixture of classic *fin de siècle* districts, with the combination of shops, offices and apartments. It was a plan which included the individualistic hallmark of architects in order to leave inhabitants with the impression of unique residential buildings. It also included creating different districts which subvert the threat of monotonous and uniform patterns. It was a plan to produce a new city centre which invites one to stroll around and linger at one's leisure. It offered free open spaces comprised of squares, promenades and terraces designed to do justice to the significance of public spaces for the quality of life and enjoyment in city districts. It also accommodated special places intended to make the HafenCity an attractive place not only for its prospective residents, but for citizens of Hamburg in general and tourists in the city in particular. Examples for such places include the Elbe Philharmonic, the Science Centre with its science theatre and aquarium, the International Maritime Museum, the Cruise Ship Terminal and, last but not least, the HafenCity University. Everything has been perfectly planned, right down to the innovative technology with which the lawns are equipped; everything from the quays made of Bongossi wood to the street lamps designed as harbour cranes has been completely stylised.

Should one compare these pawns of the master plan with the specific cultural qualities of the city as laid out by Karl Schwarz, one can say that the elements of diversity, of staging and particularly of refinement were indeed accommodated in terms of planning, but in a very specific as well as socially, culturally and economically select manner. What is missing is the element of contradiction, so that in the end, one is left with the paradox of mono-cultural diversity or, in other words, a clean, sterile diversity void of any destructive elements. What is missing are sources of friction in both a literal as well as a metaphorical sense: there is no contradiction between co-existence and incompatibility manifested in the synchronicity of all that is out of sync. What is missing are disparate, conflicting spaces and paths which one can wander from, as people say. In a place where nothing is left to chance, is there any hope for chance meetings and chance discoveries at all? Where is the real or symbolic *back alley* which belongs to each thoroughfare? Where is the backstage that belongs to each theatre? What is the essence of a place? This is a question which can be replaced by another with the help of anthropologist Pierre Sansot: '*Qu'en peut-on rever?*' – Can one dream of it? (Sansot 2004).

Karl Schwarz rightly attached a particular value to hot spots and secret places of social and cultural experiments which turned the city into a laboratory for new inventions of everything from the smallest entity of a simple design to the greatest entity of a person's life story. Such secret places have not been provided for and indeed, logically speaking, cannot possibly be provided for given that, as secrets, they cannot be planned; they simply appear at will. This clearly reveals the inherent limits of planning which show themselves at the threshold where the conceived space of the planners and the architects passes over into the inhabited and experienced space of its users. Fuelled by a misinterpretation of Richard Florida's 'Bohemian Index' (2002), ideas for planning have gone so far as to want to automatically provide for deviation. From now on, it is that particular cultural formation which has been characterised throughout history by pronounced anti-bourgeois philosophies and behaviour, namely that of the bohemian, which should serve as the catalyst for new industries and technology.

It was not by chance that the Chicago urban sociologists spoke of 'natural areas' in order to describe urban spaces which evolved in a quasi-natural, lively manner in contradiction to the administrative planning regulations. Those Chicago sociologists who had indeed become the guiding lights of urban development strategies with their perimeters of concentric circles were astonishingly sceptical with regard to the effectiveness of the planning regulations. In reference to these guidelines, Robert Park continuously emphasised the significance of the living space. 'The city, particularly the modern American city', commented Robert Park as far back as 1925,

> strikes one at first blush as so little a product of the artless processes of nature and growth, that it is difficult to recognize it as a living entity. The ground plan of most American cities, for example, is a checker-board. The unit of distance is the block. This geometrical form suggests that the city is a purely artificial construction which might conceivably be taken apart and put together again, like a house of blocks. The fact is, however, that the city is rooted in the habits and customs of the people who inhabit it. (Park 1967 [1925]: 4)

Even though the planning ideal of the chequerboard could never have made inroads in Germany, and even though, as in the case of the HafenCity, snaking lines, curves and elevations are envisaged, these merely remain suggestions which are not immune to reassignment and new uses. The rule of thumb could read as follows: the users will leave traces of their behaviour behind or else the area will be dead. 'In the course of time', continues Park, 'every section and quarter of the city takes on something of the character and qualities of its inhabitants. Each separate part of the city is inevitably stained with the peculiar sentiments of its population' (1967 [1925]: 6). What does this mean should the HafenCity actually end up with a mono-culture of wealthy people, singles and childless couples, as predicted in the *Spiegel*? What is alluded to in Robert Park's quote is all that cannot be planned, cannot be predicted in a living space; he refers to the quality of the vitalising process that lends character to a milieu that creates the special quality of the area's appearance. Within this process, planning has an accompanying and supportive function. It can only help to ensure that desired qualities are not distorted from the outset. Everything else will fall into place with time. 'In 2015, we will see if it's working' is what the Lord Mayor Ole von Beust is said to have commented on the HafenCity project.

It is the diversity of origins, practices and value systems, the contradictory ways in which they work together, alongside each other and against each other which provides the raw material for individual life stories along with the breeding ground for social visions, thereby supplying the cultural collateral of the city. It is this cultural collateral which makes cities so attractive for people who are looking to create their own unique life story beyond standard social conventions and biographical processes. It is this cultural collateral we have to thank for the creativity and innovativeness of the city which results from competition, conflict and the interplay between specialised agents generating urban synergy effects, new ideas, new trends and new lifestyles. Seen from this point of view, the city is a veritable and unique cultural and social laboratory. With regard to diversity, each city that aspires to be in this group must distinguish itself in terms of culture, lifestyle and milieus. However, as should have become clear by this stage, this achievement is not something one can 'plan'. What we can do with the help of a policy of diversity is create and preserve an environment which is necessary in order for diversity to be able to prosper. Before this can be achieved, it would surely be necessary to introduce concrete political measures such as policies of anti-discrimination and equality, smoother facilitation of naturalisation as well as improved opportunities to access education, training and employment. Another essential component over and above these measures, however, is the principle of the 'open city' which also includes the tolerance of disturbing elements, be they social outsiders or cultural strangers. It is a mistake to presume that one can rid oneself of these elements and at the same time hold on to the renewing impulses of the urban lifestyle. The spaces of freedom which are afforded by the city are inseparable. Tolerance knows only one limit and that is in relation to those who are intolerant.

In Summary

The culture of the urban lifestyle is synonymous with openness in the sense of impartiality as well as accessibility, in the sense of the undecided as well as the incomplete, in the sense of experimentation as well as the unpredictable. It is this openness which provides the soil necessary for the distinctive culture of a city to flourish, for the diversity in social worlds, cultural scenes and moral milieus. All of this, however, does not develop within a vacuum, but within a culturally codified space which is drenched in history. That is why, regardless of how often it may be suggested, there is no 'best way' that is independent of location, no magic formula for the future design of cities. A city is an individual entity with a particular biography (or history); it is a place which has been formed by its history. It is an individual entity with its own memory and its own set of habits (Lindner 2003). This has to be carefully considered in planning the future.

References

Barth, G. 1982. *City People. The Rise of Modern City Culture in Nineteenth-Century America*. Oxford: Oxford University Press.

Cox, T. 1993. *Cultural Diversity in Organisations: Theory, Research and Practice*. San Francisco: Berrett-Koehler.

Fisher, P. 1975. City Matters: City Minds, in *The Worlds of Victorian Fiction*, edited by J.H. Buckley. Cambridge, MA and London: Harvard University Press, 371–89.

Florida, R. 2002. *The Rise of the Creative Class: And How it's Transforming Work, Leisure, Community and Everyday Life*. New York: Basic Books.

Führer, K.C. 2006. Stadtraum und Massenmedien, in *Zentralität und Raumgefüge der Großstädte im 20. Jahrhundert*, edited by C. Zimmermann. Stuttgart: Steiner, 105–20.

Hannerz, U. 1980. *Exploring the City. Inquiries Toward an Urban Anthropology*. New York: Columbia University Press.

Kroeber, A.L. 1948. *Anthropology*. New York: Harcourt, Brace & World.

Lindner, R. 2003. Der Habitus der Stadt – ein kulturgeographischer Versuch. *Petermanns Geographische Mitteilungen* 147: 46–53.

Lindner, R. 2005. Die Kultur der Metropole. *Humboldt Spektrum* 12(2): 22–8.

Lofland, L. 1973. *A World of Strangers*. New York: Basic Books.

Park, R.E. 1915. The City: Suggestions for the Investigation of Human Behavior in the City Environment. *American Journal of Sociology* 20(5): 577–612.

Park, R.E. 1967 (1925). The City: Suggestions for the Investigation of Human Behavior in the Urban Environment, in *The City*, edited by R.E. Park and E.W. Burgess. Chicago and London: University of Chicago Press, 1–46.

Sansot, P. 2004. *Poétique de la ville*. Paris: Payot and Rivages.

Schwarz, K. 1987. Die Metropolen wollen. Berlin als Metropole wollen, in *Die Zukunft der Metropolen: Paris, London, New York, Berlin*, edited by K. Schwarz. Berlin: Technische Universität Berlin, 21–30.

Simmel, G. 1903. Die Großstädte und das Geistesleben, in *Die Großstadt. Vorträge und Aufsätze zur Städteausstellung*, edited by T. Petermann. Dresden: Gehe-Stiftung zu Dresden, 185–206.

Stuart, J. 1972. Introduction. The Education of John Reed, in *The Education of John Reed. Selected Writings*, edited by J. Stuart. Berlin: Seven Seas Publication, 9–25.

Index